Lecture Notes in Engineering

Edited by C. A. Brebbia and S. A. Orszag

DLR 60

U. Schumann (Editor)

Air Traffic and the Environment – Background, Tendencies and Potential Global Atmospheric Effects

Proceedings of a DLR International Colloquium
Bonn, Germany, November 15/16, 1990

Springer-Verlag Berlin Heidelberg GmbH

Editor
U. Schumann
Deutsche Forschungsanstalt
für Luft- und Raumfahrt (DLR)
Institut für Physik der Atmosphäre
8031 Oberpfaffenhofen
Germany

ISBN 978-3-540-53352-8 ISBN 978-3-642-51686-3 (eBook)
DOI 10.1007/978-3-642-51686-3

Originally published by Springer-Verlag Berlin Heidelberg New York in 1990.

2161/3020-543210 Printed on acid-free paper.

PREFACE

The strong and continuing increase in air traffic, the plans to build supersonic aircrafts and hypersonic space-transport systems, the developments of hydrogen technology, and the general concern on global changes have raised questions on effects of emissions from air traffic on the environment and especially the atmosphere above and shortly below the tropopause. What are the consequences of water vapour emissions on the formation of high clouds and global climate? What are the possible effects of emissions on the ozone layer in the stratosphere and upper troposphere? Which technological developments can help to reduce the emissions?

These questions get increasing attention in the public. Some previous meetings of scientific experts have shown that the topic is of high interest but most questions cannot be answered yet to a sufficient degree. More research is necessary and the topic requires interdisciplinary cooperation. Moreover, there is a need to document the basic knowledge required to assess possible consequences of increasing and changing traffic. With respect to possible global changes, air traffic at cruising altitude seems to have the most important influence and it becomes necessary to consider technological alternatives.

The German Aerospace Research Establishment (DLR) has initiated a series of seminars on fundamental problems of sciences in which DLR is envolved. Previous seminars considered:

1984 Nonlinear Dynamics of Transcritical Flows
1985 Uncertainty and Control
1986 Artificial Intelligence and Man-Machine-Systems
1987 Parallel Computing in Science and Engineering
1988 Hydrocarbon Oxidation
1989 Optimization, Methods and Applications, Possibilities and Limitations

This book contains ten papers which had been prepared for presentation at the 1990 DLR-Seminar on

Air Traffic and the Environment -
Background, Tendencies and Potential Global Atmospheric Effects.

At the seminar, an additional paper is to be presented by Dr. Dieter H. Ehhalt. He will talk about "The background distribution of NO_x and hydrocarbons in the upper troposphere." Some of his results are published in a paper by him with J. W. Drummond in "Tropospheric Ozone" (ed. I. S. A. Isaksen, ed.), p. 217-237, D. Reidel Publ. Co. (1988). I had also invited Dr. Andy Heymsfield, National Center for Atmospheric Research, Boulder, Colorado to present his recent results from a model of contrail formation and comparisons to experimental data. Unfortunately, he cannot be at the Seminar because he is participating in a field measurement program at the time of the Seminar and unable to take the time off.

As in previous cases, the papers to be presented at the seminar have been collected before the seminar takes place. As a consequence, the oral presentations might contain even more recent results. Moreover, there was practically no time to revise or edit the text. These drawbacks of a pre-seminar publication should be balanced by the more precise information of the seminar participants and early publication of the results.

The purpose of the papers is to describe the state of the knowledge, new facts and findings and open questions. The presentations during the seminar should result in interdisciplinary discussions and interactions. It might initiate further research projects and contribute to appropriate actions.

The papers present facts and research results concerning

- The development of air traffic and fuel consumption, with special emphasis to long-distance traffic and high flying aircrafts,
- Technological developments which may change emissions, including liquid hydrogen propulsion systems,
- the emissions from air traffic and potential developments in turbine-engine technology to reduce such emissions,
- background information on the global atmospheric circulation and transport affecting the life-time of air constituents, and on typical background concentration profiles,
- consequences for the ozone layer, e.g. due to changing air chemistry by aircraft emissions,
- and consequences for the greenhouse effect, e.g. due to water-vapour emissions, and their potential impact on the formation of contrails and extended cirrus clouds.

The papers have been written by internationally recognized experts in the respective fields. I wish to express my thanks to all authors who contributed to this publication. I am also grateful to the publishers for their outstanding cooperation in the printing of this book.

Ulrich Schumann
Oberpfaffenhofen, August 1990

CONTRIBUTORS

Dr. Christoph Brühl
Max-Planck-Institut für Chemie
Abteilung Chemie der Atmosphäre
Universität Mainz
Postfach 3060
6500 Mainz, Germany

Prof. Dr. Paul Crutzen
Max-Planck-Institut für Chemie
Abteilung Chemie der Atmosphäre
Universität Mainz
Postfach 3060
6500 Mainz, Germany

Prof. Dr. Peter Fabian
Institut für Bioklimatologie
und Angewandte Meteorologie
Universität München
Amalienstr. 52
8000 München 40, Germany

Prof. Dr. Hartmut Graßl
Max-Planck-Institut für Meteorologie
Universität Hamburg
Bundesstr. 55
2000 Hamburg 13, Germany

Dr. Hubert Grieb
Motoren und Turbinen Union München
Postfach 500640
8000 München 50, Germany

Dr. Douglas E. Kinnison
Atmospheric and Geophysical Sciences Div.
Lawrence Livermore Laboratory
P.O. Box 808, L-262
Livermore, California 94550, USA

Dr. George G. Koenig
Geophysics Laboratory
Hanscom Air Force Base
Bedford, Massachusetts 01739, USA

Prof. Dr. Kuo-Nan Liou
Department of Meteorology
University of Utah
Salt Lake City, Utah 84112, USA

Dr. Hans-Gustav Nüßer
Deutsche Forschungsanstalt für
Luft- und Raumfahrt
Hauptabteilung Verkehrsforschung
Postfach 906058
5000 Köln 90, Germany

Dr. Szu-Cheng S. Ou
Department of Meteorology
University of Utah
Salt Lake City, Utah 84112, USA

Dipl.-Ing. Hans-Peter Reichow
Deutsche Lufthansa AG, HAM CU
Postfach 630 300
2000 Hamburg 63, Germany

Dr. Alfons Schmitt
Deutsche Forschungsanstalt für
Luft- und Raumfahrt
Hauptabteilung Verkehrsforschung
Postfach 906058
5000 Köln 90, Germany

Prof. Dr. Ulrich Schumann
Deutsche Forschungsanstalt für
Luft- und Raumfahrt,
Institut für Physik der Atmosphäre
8031 Oberpfaffenhofen, Germany

Dr. Burkhard Simon
Motoren und Turbinen Union München
Postfach 500640
8000 München 50, Germany

Dr. Peter Wendling
Deutsche Forschungsanstalt für
Luft- und Raumfahrt,
Institut für Physik der Atmosphäre
8031 Oberpfaffenhofen, Germany

Prof. Dr. Carl-Jochen Winter
Deutsche Forschungsanstalt für
Luft- und Raumfahrt,
Pfaffenwaldring 38-40
7000 Stuttgart 80, Germany

Dr. Donald J. Wuebbles
Atmospheric and Geophysical Sciences Div.
Lawrence Livermore Laboratory
P.O. Box 808, L-262
Livermore, California 94550 USA

CONTENTS

THE GLOBAL DISTRIBUTION OF AIR TRAFFIC AT HIGH ALTITUDES, RELATED FUEL CONSUMPTION AND TRENDS

H.-G. Nüßer, A. Schmitt

DLR, Transport Research Division
D-5000 Köln 90, Federal Republic of Germany

ABSTRACT

The total air transport system consists of many components, the main parts being the military and the civil aviation which consists again of the commercial air transportation with scheduled and charter air services and of the non-commercial or general aviation. Whereas general aviation can be regarded as a means of flying primarily for private and leisure purposes of individuals in lower altitudes - however, business travel by company owned aircraft belongs to general aviation as well - it is the commercial air transportation flying in high altitudes which serves the general public as a transport mode providing national and international services.

This paper is concerned with the analysis, forecast and global distribution of the commercial air traffic, gives a first rough estimate of global fuel consumption and aircraft emitted pollutants, and describes the method of assigning aircraft movements to global traffic cells as the basis for estimating the pollution caused by these movements.

INTRODUCTION

Hand in hand with economic growth, growing importance and availability of leisure time, and with technological progress in aviation demand for air transportation has grown considerably world wide. Corresponding to ICAO, "in 1988 the number of passengers carried yearly by the world's airlines approximated one fifth of the world's population. Almost a quarter (by value) of world trade in manufactured goods is now carried by air and the portion is growing. Airline fleets grew significantly in parallel with continuing traffic growth."

Although the share of air transportation in the total transport demand is still relatively small, there are some reasons to take the emission of air pollutants by aircraft seriously. The fuel consumption of air traffic has reached meanwhile a high level, in the higher parts of the atmosphere pollution comes from air traffic only, in altitudes above the tropopause pollutants are active for a long time, and one has to expect a further increase of fuel consumption parallel to the growth of demand.

In this paper, a first attempt has been made to estimate the global distribution of air traffic and of fuel consumption on the basis of ICAO statistics and Lufthansa data. Unfortunately the data tape with the statistics of international air traffic movements between airports by aircraft type has not yet been made available, so that this paper describes only the method of assignment. With presentation of the paper in November, there will be available an assignment of air traffic for the year 1988 to global cells.

GLOBAL AIR TRANSPORT DEMAND

As shown in **Fig. 1** global air traffic of scheduled services has grown from 940 billion passenger-km in 1978 to almost 1 700 billion passenger-km in 1988, 45% of the total being international traffic. The annual growth rate in the decade between 1978 und 1988 was 6.1% and the ICAO forecast expects nearly the same rate for the next 12 years up to the year 2000, i.e. a doubling of the total transport performance.

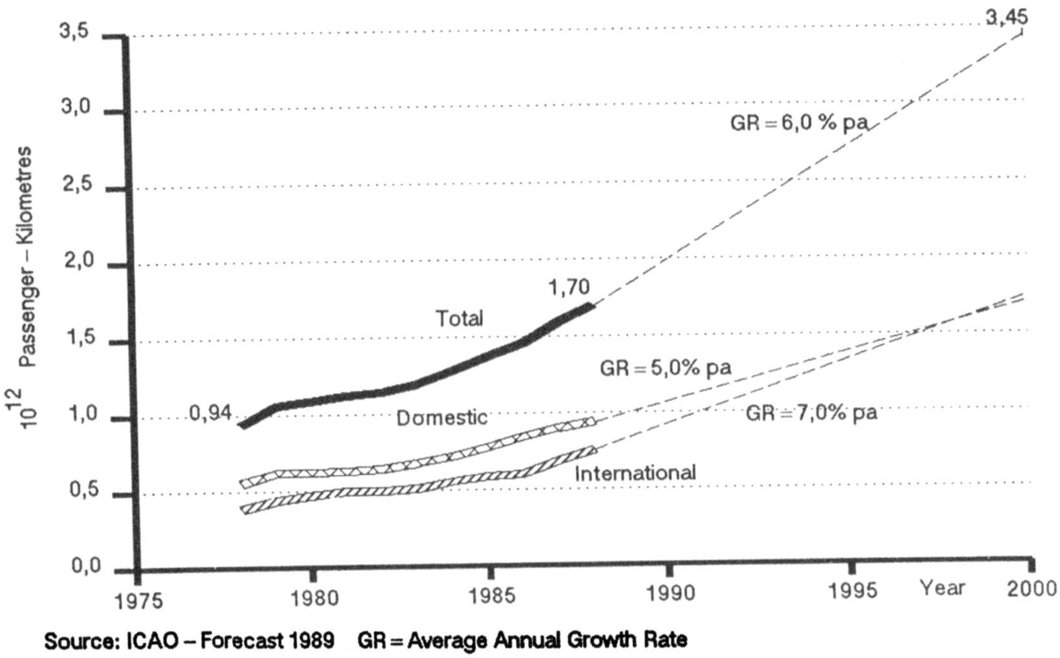

Fig. 1: World Scheduled Passenger Transport Demand

In some markets of world wide air transportation scheduled operations are supplemented by charter services, serving primarily holiday travellers. Total international passenger-kms performed in non-scheduled traffic throughout the world (see **Fig. 2**) reached in 1988 a volume of about 170 billion, thus representing almost 19% of total international travel on scheduled and charter services. Travel between the mediterranean Region and European Countries constitutes the world's largest charter market, the volume of non-scheduled traffic being comparable to that of scheduled traffic. Since charter traffic did not succeed in gaining world wide coverage it did not grow as fast as scheduled services; in the decade 1978-1988 the average annual growth rate of passenger-kms performed in charter transport was 4.3%.

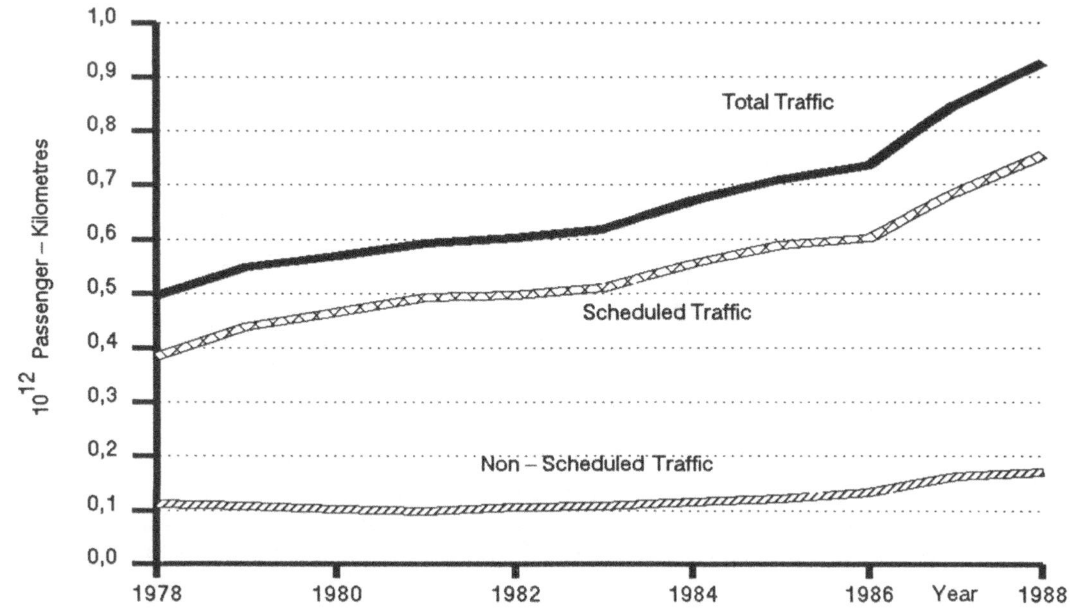

Source: ICAO

Fig. 2: World International Scheduled and Non-Scheduled Passenger Transport Demand

Global demand figures are primarily of statistical value; for analysing and forecasting air traffic, structural characteristics like regional, relational, or travel purpose data have more significance. Regional or national markets vary widely in size, travel structure, network density and traffic development patterns. **Fig. 3** shows the past development and forecast of passenger-km performed by scheduled services of airlines registered in ICAO states of each region. In 1988 about 42% of the total passenger-km volume handled by airlines of almost 160 Contracting States of ICAO is accounted for the most important air traffic region North America, 30% for Europe, and only 28% for all other regions. However, regional air traffic growth was strongest in Asia and the Pacific, with its traffic share increasing from 13.4% in 1978 to 18.2% in 1988, and which is expected to increase to 26.1% in 2000, with an average annual growth rate of about 9.5%, compared to the rate of total traffic of 6%.

For the first time, the total commercial air transport fleet of all scheduled and non-scheduled carriers of ICAO Contracting States (except China and USSR) exceed the number of 10 000 aircraft in 1987. There have been important changes in the composition of airline fleets in the past. The number of jet aircraft (see **Fig. 4**, sum of all wide body and narrow body) increased from 5684 in 1978 to 8086 in 1988, rising from 68% to 76% of the fleet. While the number of turboprop aircraft increased as well (1416 to 1910) the number of piston engine aircraft dropped from 1269 to 590 in the same decade, constituting only about 5.5% ot the total fleet. **Fig. 4** also shows clearly the trend of airlines to buy 2 engine jet aircraft instead of 4 engine aircraft especially for intercontinental flights.

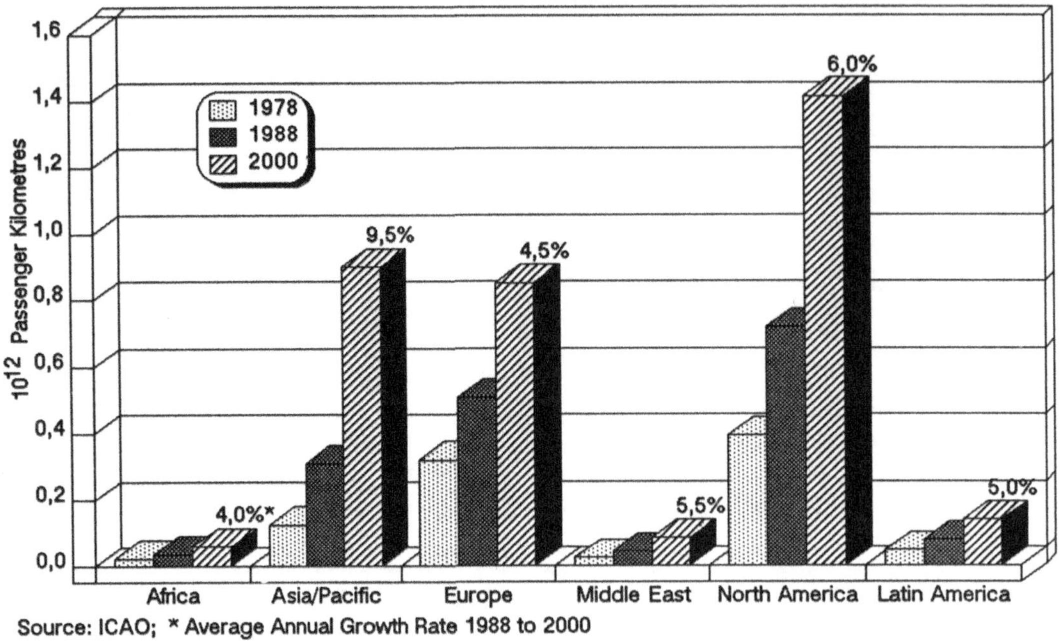

Fig. 3: Forecast of World Scheduled Passenger Traffic by Region

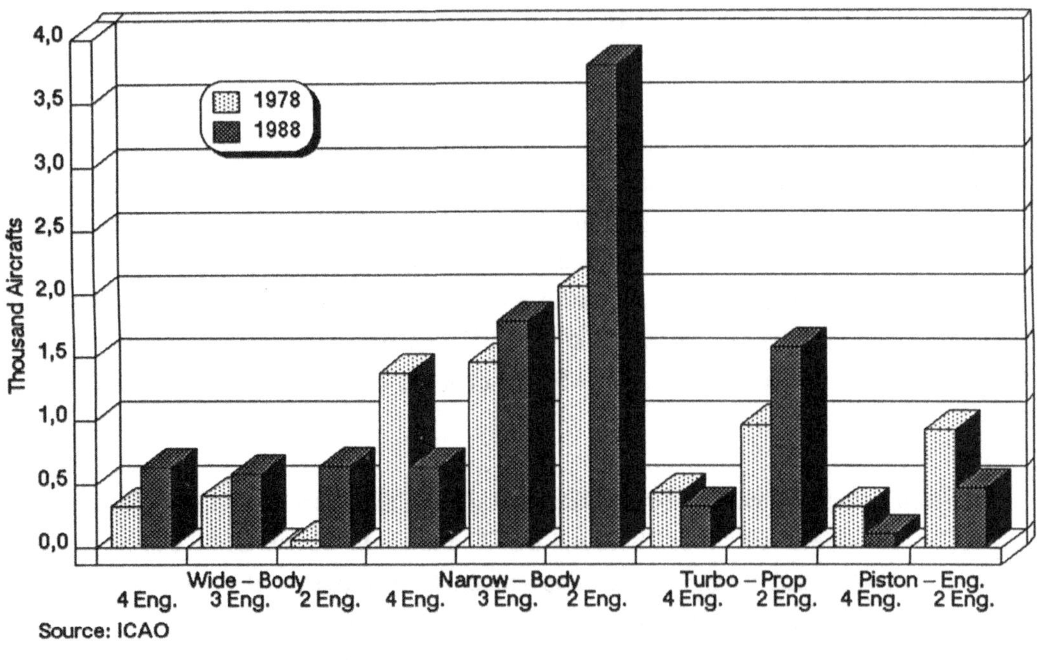

Fig. 4: Total Commercial Transport Fleet Distinguished by Type of Propulsion

Finally a comparison of different forecasts has been made in order to show the expectations of the air industry regarding the future growth of air transportation.

Source	Range of Forecast	Passenger (Pkm) GR (%)	GF	Freight (tkm) GR (%)	GF
ICAO (1989)	1988-2000	6.0	2.0	7.0	2.3
IATA (Sept. 1989)[1)]	1988-1993	7.05	1.4	7.0	1.4
Airbus (Nov. 1987)	1986-1996	5.7	1.74	6.7	1.9
	1996-2006	5.3	1.68		
	1986-2006	5.5	2.92		
Boeing (Febr. 1989)	1987-2000	5.9	2.1	6.0	2.1
	2000-2005	4.2	1.23		
	1987-2005	5.4	2.6		
McDonnel Douglas (Febr. 1989)	1987-2002	5.7	2.3	6.0	2.4
MBB (Juni 1989)	1988-2008	5.12)	2.7		

Remarks: 1) Passengers and Freight in Border-Crossing Traffic
2) Scheduled and Charter Services
GR = Average Annual Growth Rate; GF = Growth Factor

Source: Wilken [6]

Tab. 1: Comparison of Forecasts of Global Air Transport Demand - Scheduled Services

Table 1 shows the available forecasts of global air transport demand for scheduled services. Institutions like ICAO and IATA, aircraft manufacturers like Airbus, Boeing, McDonald, MBB estimate the further development by means of an econometric function which gives the air transport demand in passenger-kms performed depending on the Gross Domestic Product and Yield (mean revenue per passenger-km). Besides all differences in the forecasts there are similar opinions that:

- world wide air transport demand measured in passenger-km performed will double between 1988 and 2000,
- air transport performance in freight traffic will increase stronger than passenger traffic,
- there is no indication of saturation in air transport demand after the year 2000.

FUEL CONSUMPTION AND AIRCRAFT EMITTED POLLUTANTS

Although the share of air transportation of the total transport demand is still relatively small, there are some reasons to take the emissions of air pollutants by aircrafts seriously.

- In absolute terms fuel consumption by air traffic has meanwhile reached a high level. IATA has estimated the total fuel consumption of global civil aviation to about

$120*10^6$ tons for the year 1987; the UN Yearbook gives $107*10^6$ tons for the same year; DLR calculations on the basis of ICAO and Lufthansa figures come up with $150*10^6$ tons in 1988.

- In higher parts of the atmosphere pollution by air traffic can be comparably strong, even dominant. An important amount of kerosine is consumed when flying in or above the tropopause - in case of Lufthansa about 20%. (The altitude of the tropopause, which separates the troposphere from the stratosphere, grows with descending latitude from 8 km in polar regions to 16 km in the near of the equator, depending also on season and weather conditions.)

- Only a small percentage of the pollutants emitted by ground traffic passes the troposphere. That means that in the troposphere aviation is the main source of pollutant emission caused by traffic.

- Supersonic aircrafts like "Sänger" would fly as high as 36 km. In this altitude their emissions would be the only anthropic source of pollution.

- Moreover, because the natural background concentration of water vapor, nitrogen oxids and hydrocarbons is low above the tropopause, in these altitudes anthropic emissions of such gases can easily reach comparable concentrations.

- In altitudes above the tropopause pollutants are active for a long time. In the troposphere atmospheric constituents are mixed relatively quickly and intensively in vertical direction and trace gases as air pollutants are typically washed out within less than a week, e.g. by precipations formed in that layer. Above the tropopause temperature is ascending with altitude and a relatively stable stratification is formed, in which trace gases can remain as long as one year or more in higher altitudes.

The emission of the pollutants is depending on many different factors. Some of the underlying processes are far from being satisfactorily understood. Essentially, emissions of water vapor (H_2O) and carbon dioxide (CO_2) are proportional to fuel consumption. The amount of unburned hydrocarbons (C_xH_y) and the formation of carbon monoxide (CO), nitrogen oxides (N_xO_y), and particulate matter (C) depends in the main on engine type, power setting, flight altitude, and air speed. The volume of sulphur dioxide (SO_2) emission is mainly determined by fuel quality.
Table 2 shows the specific emission of these pollutants exhausted by aircraft under cruise conditions in g/kg Kerosine, as estimated by Lufthansa for its fleet.

	[g/kg]	Mainly depending on
H_2O	1239	fuel consumption
CO_2	3154	fuel consumption
CO	0.7 - 2.5	engine type
N_xO_y	6.0 - 16.4	power setting
C_xH_y	0.05 - 0.7	flight altitude
(C)	0.007 - 0.03	airspeed
SO_2	1	fuel quality

Source: Lufthansa, 1989

Tab. 2: Specific emissions of air pollutants under cruise conditions for the Lufthansa fleet

The Lufthansa data include values for fuel consumption and available t-kms by aircraft type, that is aircraft with short and long ranges. Combining this information with ICAO data on world wide available t-kms leads to a first estimation of the fuel consumption in world scheduled domestic and international passenger traffic in the year 1988, in order to get a first assessment of worldwide quantities of aircraft-emitted pollutants. The result of this calculation is shown in **Table 3**.

		Domestic		International		Total	
		1988	2000	1988	2000	1988	2000
Traffic	(10^9 Pkm)[1)]	940	1710	756	1740	1696	3450
Fuel	(10^6 t)	105	190	45	110	150	300
CO_2	(10^6 t)	325	590	145	330	470	920
H_2O	(10^6 t)	130	230	60	130	190	360
C	(10^3 t)	2.1	3.7	0.5	1.0	2.6	4.7
C_xH_y	(10^3 t)	60	105	30	70	90	175
CO	(10^3 t)	225	410	45	105	270	515
N_xO_y	(10^3 t)	785	1430	730	1680	1515	3110
SO_2	(10^3 t)	105	190	45	110	150	300

1) World Scheduled Passenger Traffic including USSR, Source: ICAO

Tab. 3: Fuel Consumption and Air Traffic Pollution for the Scheduled Air Traffic

On average the Lufthansa fleet consists of more modern aircraft than the world aircraft fleet operating in scheduled air traffic. Consequently, on the average, fuel consumption and specific pollutant emission are comparably lower than for the world fleet. Therefore the fuel consumption has been increased by 20% in the calculation.

Following these results, the total amount of CO_2 and H_2O exhausted by scheduled aviation is in the dimension of some hundreds of million tons, the emission of N_xO_y is about 1,5 million tons, the emission of C_xH_y, SO_2, and CO in the dimension of 100 to 300 of thousand tons, and the emission of particulate matter about 3 thousand tons.

If no efforts are made to reduce fuel consumption and pollutant emission the quantities will nearly double until the year 2000. (See **Table 3**, last column.)

GLOBAL DISTRIBUTION OF AIR POLLUTANTS EMITTED BY AIRCRAFT

Within the DLR-program "Air Traffic and Environment" the Transport Research Division of the DLR has untertaken the task to develop a first idea about, where pollutants exhausted by aircrafts are emitted around the globe.

To achieve this objective the ICAO data of the "Traffic by Flight Stage" [3] will be used as the main source of information. It contains data on traffic by flight stage operated on scheduled international services performed during the year 1988. For 662 airports, 97 airlines (using 107 different types of aircrafts) of 58 Contracting States of ICAO data are available on data tape. To calculate the global emission, data of the German airline Lufthansa [5] on fuel consumption and on emissions of air pollutants of the Lufthansa fleet (7 aircraft types with ranges from 565 km to 5100 km) during the year 1988 are available.

In order to deduce an algorithm for a rough estimation of the global distribution of where the pollutants emitted by aircraft are exhausted, a model with simplifying assumptions has to be defined:

- The earth is regarded as a regular sphere (with radius $R = 6371,221$ km), for which the formulas of the spherical trigonometry are valid.

- Flights from airport A to airport B and vice versa are characterized by the following features

  -- The shortest connection on the great circle is used,

  -- For the total flight distance a flight under cruise conditions is assumed for the calculation (with a flight altitude of 10 km and higher).

  The first assumption leads to an underestimation, the second to an overestimation of fuel consumption and the deduced volumes of emissions.

- Specific fuel consumption and specific emission of pollutants should be related to engine types. However the ICAO statistics used include only aircraft-related data, and there is probably no chance to get more information which would one enable to assign correctly aircraft to engine type. Therefore, we have to use aircraft-related values for specific fuel consumption and specific emissions.

For estimating the spatial distribution of pollution the globe has been divided into 648 cells of equal area ($787*10^3$ km^2). The area of each cell is defined by two pairs of respective geografical latitudes and longitudes (see **Table 4**). (A cell definition based on equal length of steps for both the geografical latitude and altitude would result in very different cell areas. With steps of respective 10° the smallest cells near the poles would have $107*10^3$ km^2, the largest-ones in the near of the equator $1233*10^3$ km^2).

Northern Hemisphere

Cell No.	Geographical Latitude [Φ]	Geographical Longitude [l]
1 - 36	90.0000°N ≤ Φ ≤ 62.6746°N	Subdivision of
37 - 72	62.6746°N < Φ ≤ 50.9863°N	each class of
73 - 108	50.9863°N < Φ ≤ 41.7379°N	36 cells by steps
109 - 144	41.7379°N < Φ ≤ 33.6817°N	of 10°,
145 - 180	33.6817°N < Φ ≤ 26.3298°N	starting with
181 - 216	26.3298°N < Φ ≤ 19.4255°N	0° ≤ l < 10°W
217 - 252	19.4255°N < Φ ≤ 12.8080°N	and proceeding
253 - 288	12.8080°N < Φ ≤ 6.3633°N	to the West
289 - 324	6.3633°N < Φ ≤ 0.0°	

Southern Hemisphere

Analogous method for the definition of the cells No. 325 - 624 of the southern hemisphere, with numeration starting from the equator.

Tab. 4: Definition of 648 Globe Cells

The aim is, to assign to these cells a corresponding part of the global emission of pollutants exhausted by aircraft. To get this distribution the following steps are carried out:

(a) Calculation of d_{AB}, the shortest distance following the great circle between airport A and airport B for each airport-pair of the ICAO statistic used, with respective geographical latitudes Φ_A, Φ_B and geographical longitudes l_A, l_B.

(b) Calculation of the fraction $d(Z)_{AB}$ of the total flight distance d_{AB}, which has to be assigned to each cell Z passed on the flight.

(c) For all types of aircraft T used for flights between airport A and airport B, determination of the annual amount of kilometers $P(Z,T)_{AB}$, flown when passing the area of cell Z, taking into account the number of annual flights $F(T)_{AB}$ between airport A and airport B with aircraft of type T.

(d) Summation of the aircraft-type-specific annual amount of kilometers $P(Z,T)_{AB}$, assigned to cell Z, over all connections between the 662 airports of the ICAO statistics. (The result is called P(Z,T)).

(e) Estimation of the annual fuel consumption V(Z,T) of all aircraft of type T, which has to be assigned to cell Z, taking into account the specific fuel consumption v(T) of an aircraft of type T (equipped with a known engine type).

(f) Estimation of the annual volume of emission of pollutant i, named $E_i(Z,T)$, exhausted by the total number of aircraft of type T when passing the area of cell Z, taking into account the rate of emission for pollutant i per kg of fuel consumed ($e_i(T)$), specific for an aircraft of type T (equipped with a special, known type of engine).

(g) Summation over all 107 types of aircraft of the ICAO statistics to get $E_i(Z)$, which is the total emission for each pollutant i that has to be assigned to cell Z, considering all flights of the ICAO statstics.

(h) Using the ICAO statistics only a part of all flights is taken into consideration. The ICAO statistics do not contain:

- military aviation
- charter air services
- scheduled domestic flights
- international flights of airlines for which data are not available for the year 1988 (97 of the 212 airlines of the Contracting States of ICAO have reported statistics to ICAO, fortunately most of the important-ones are included).

To improve the results based on the used ICAO statistics, available complementary information will be taken into consideration. E.g. the results of step (g) can possibly be corrected by adjusting the value of sums of cell-related results to known respective regional and global data. The algorithms of the steps (a) - (g) are shown in **Table 5.**

Our outcome will be input data for other projects of the DLR-program "Air Traffic and Environment". Therefore we have harmonized our procedure to the demands of the whole program. The presentation and discussion of the results will show, regarding to which aspects more details are needed to fit additional requests.

Step		Algorithm	Unit
(a)	d_{AB}	$= \arccos(\sin\Phi_A * \sin\Phi_B + \cos\Phi_A * \cos\Phi_B * \cos(l_A - l_B))$	[km]
(b)	$d(Z)_{AB}$	calculated by means of a DLR-computer-program (using formulas of spherical trigonometry)	[km]
(c)	$P(Z,T)_{AB}$	$= F(T)_{AB} * d(Z)_{AB}$	[km/a]
(d)	$P(Z,T)$	$= \Sigma_{\text{all connections}}\ P(Z,T)_{AB}$	[km/a]
(e)	$V(Z,T)$	$= v(T) * P(Z,T)$	[t/a]
(f)	$E_i(Z,T)$	$= e_i(T) * V(Z,T)$	[t/a]
(g)	$E_i(Z)$	$= \Sigma_{\text{all types of aircrafts}}\ E_i(Z,T)$	[t/a]
(h)	corrections using additional information		

Tab. 5: Algorithm for the Estimation of the Global Distribution of Pollutants Emitted by Aircraft

REFRENCES

[1] Boeing, 1989: Current Market Outlook. World Travel Market Demand and Airplane Supply Requirements. *Boeing Commercial Airplanes*

[2] ICAO, 1989: The Economic Situation of Air Transport. Review and Outlook 1978 to the Year 2000. *ICAO Circular 222-AT/90*

[3] ICAO, 1989: Traffic by Flight Stage. *ICAO Series TF-No. 103, Digest of Statistics No. 365*

[4] Nüßer, H.-G. and Wilken, D.,1990: Status and Future Development of Air Transport Demand. *Proceedings European Propulsion Forum: Future Civil Engines and the Protection of the Atmosphere, 13-20.*

[5] Reichow, H. P. (Lufthansa), 1989: Der Luftverkehr aus der Sicht eines Nutzers. *In proceedings Expertentagung "Energie - Emission -Klima: Lösungsansätze für den Verkehr*

[6] Wilken, D., 1990: Luftverkehrsprognosen: Problematik, Vergleich und Ergebnisse. *Lufthansa Jahrbuch '90, 150-158*

Fuel Consumption and Emissions of Air Traffic

Hans-Peter Reichow

Lufthansa German Airlines

Commercial air traffic worldwide is growing faster than the world economy. Trade and air transport are important tools for the international division of labour. Worldwide communication has shortened distances, and it actually has created - contrary to some previous prognoses - additional demand for travel and transportation. Political development between East and West is further accelerating growth in world trade. Advancing liberalization in Eastern Europe is giving impetus to considerable economic growth over the medium term which will create increased demand for air transport services, especially in countries whose highway and railroad systems are yet inadequate.

The free market is replacing tariffs approved by the authorities and is putting pressure on prices. This tendency will intensify as the EG domestic market develops. Air travel is no longer a privilege for the rich. In comparison with the general cost of living, flying has become cheaper. The numbers of people on private travel is already greater than those on business travel, and this proportion is increasing.

Considering all this, the following prognosis is perfectly reasonable: Air transport worldwide will double by the year 2000. (Graphic: World scheduled air traffic).

In line with Germany's economic status and its geographic location, Lufthansa´s share in this growth will be above average. Freedom of establishment and sharper competition in Europe will mean that smaller airlines will have a difficult time surviving without partners. Larger carriers, by comparison, will be able to expand their markets through cooperation.

A new commercial aviation hub in central Europe will develop, which might help to relieve some airports in West Germany whose capacities are close to saturation.

Anticipating this projected growth, our responsibility for the environment is vital. Developing appropriate strategies requires precise data, reliable scientific information, international consensus, and dependable, steady, legislation. In addition, we intend - in carrying on with our fleet policy - to continue promoting the development and procurement of environment-friendly aircraft, while not losing sight of economic common sense. (3 graphics: Fleet Renewal, Age of Lufthansa Fleet, Fuel Consumption 1970-1989)

We are also committed - as in the past 30 years - to promoting the introduction of suitable operating procedures. However, there is no sense in national, or even European, isolated actions. Good will alone can not help the environment, if pushed out of the market.

In June last year, we provided the German Bundestag Commission for Protection of the Earth Atmosphere with complete information about performance, fuel consumption and emissions of the Lufthansa fleet in 1988. Some of you here will be aware of this information. I would like to show you the updated set of data for 1989. (Graphics 1, 2, 3)

The figures on fuel consumption and emissions in passenger and freight services need clear definition because all passenger aircraft carry also cargo, and because the revenue load may be limited by weight or by volume, depending on aircraft version and route.

The P = 2 factor assigns passenger tonne-kilometres double the fuel consumption over freight tonne-kilometres in a given airplane with a given load and a given consumption. This factor takes into account the additional weight involved in passenger transportation. Furthermore, in the case of the various different versions of the Boeing 747-200 (passenger, combi and freighter airplanes), this factor 2 provides practically equivalent specific consumption data for all versions.

As you can see, we need, on average, around 10 litres of kerosine per 100 person-kilometres with our short range aircraft, while the medium and long range aircraft use well under 6 litres per 100 person-kilometres. The entire Lufthansa fleet used, on average during 1989, 6.6 litres per 100 person-kilometres. (The kilometres applied here are great-circle distances - the shortest distances between take-off and landing, regardless of the actual route taken.)

Please note that the 747-400 and the A-320 fleets have only just started operation with a few airplanes in the latter part of the year, so that data are not yet typical. Due to late delivery of the 747-400, this airplane had to be used - quite untypically - on medium and short-haul routes.

The next chart (Specific Emissions) indicates the emissions of Lufthansa aircraft in terms of passenger-kilometres carried and tonne-kilometres of freight. The figures provide a realistic basis of comparison with other means of transportation. These data relate to 1988, because the 1989 analysis was not yet ready at the time of this writing. Significant differences are not expected, because the new fleets did not yet contribute much to the 1989 data.

For flights in the stratosphere (17 to 20% of the Lufthansa network), aircraft emissions account for a considerable portion of anthropogenic substances. Of particular significance here are nitrogen oxide and water. To date, though, no solid information is available regarding reaction cycles, dwell periods and means of conveyance. Related research programs are being defined by the DLR.

The suggestion that air traffic should operate at lower altitudes, as a percautionary measure, could prove fatal because fuel consumption, and thus emissions, would be certain to increase. (Graphic: Lufthansa Flights In and Above the Troposhere). In the prevailing weather of June 22, 1989, fuel consumption for typical flights increased between 6 and 8 %.

We await with great interest the results of scientific research, and, providing there is worldwide agreement, we will not hesitate supporting appropriate measures in IATA and ICAO.

Recently, we visited two major engine manufacturers in the United States. We made it urgently clear to them that reduction of specific nitrogen oxide emissions will probably be be a central goal of engine development in the coming years, and that this parameter will be a significant criterion for Lufthansa in selecting future equipment. This announcemnt from an important customer - which already has frequently influenced trend-setting investment decisions - was not without effect.

On the other hand, we fully realize that we cannot expect major success over the short-term, and also that the cost involved in developing new engines is immense. In the end, additional cost will be passed on to our customers. That is when we will see whether environment-consciousness is merely high-minded thinking or truly indicates a readiness to make sacrifices.

Short-range traffic, that is under 400 kilometres air distance, belongs on the rails, for both economic reasons and in the interest of the environment. The stipulation for this being a European high-speed rail network and efficient rail stations beneath every international airport terminal.

It is somewhat odd that, recently, environment-conscious citizens have been campaigning against such plans. Is the local restriction of a runway (comparable to about four kilometres of Autobahn) easier to accept, after all, than the linear expansion of high-speed railroad tracks? - Political decisions are required here.

One disappointing example is the new Munich 2 Airport, which will not be accessible by long-distance rail when it goes into service. It is still not clear at present whether it will be later linked to the high-speed rail network or remain accessible only by local commuter train. We plan to continue cooperating with the railway system. We have been much encouraged by the success of our Airport Express trains between Düsseldorf and Frankfurt, and between Stuttgart and Frankfurt. More destinations are to come.

The air traffic control system in Europe is outmoded. (Graphic Air Traffic Control Compared) There are 42 air traffic control centres operating 22 different systems, and this hinders optimum cooperation. In the United States there are 20 centres operating a single system. The technical and personnel capabilities of air traffic control in both Germany and the rest of Europe are inadequate. Airport capacities are inadequate. And the division of airspace between military and civilian air traffic is outdated and not sufficient for the needs of commercial aviation.

During 1989, the various shortcomings mentioned created 9200 hours of holding time for Lufthansa aircraft over Germany alone. This is politically induced pollution. Adequate measures at the European level are urgently required.

Capacity bottlenecks will not grow fewer as air traffic growth continues. The number of widebody aircraft will therefore increase substantially. There will be more non-stop flights, on both continental and intercontinental routes.

Loud aircraft will soon be a thing of the past for us. When the last Boeing 727-200s have been mustered out by 1992, and the 737-200s a year later, we will then be operating exlusively quiet aircraft meeting the requirements of Chapter 3, ICAO Annex 16. The problem of noisy aircraft will then be dormant for us - that is, until everyone else has acquired quiet aircraft also!

This diagram (Graphic Noise Contour) shows the dramatic reduction in the aircraft noise carpet achieved with the Airbus A320 compared to the Boeing 727. It shows the 85 dB(A) contour on the surface of the airport area: The Airbus carpet is completely confined to within the airport fencing. (The 85 dB(A) level is equivalent to he noise from a train heard from a distance of 50 metres.)

Night-flying restrictions on quiet aircraft in the Chapter 3 category must be lifted. On the other hand, loud aircraft could be forced out of traffic by appropriate operating restrictions and steep fees.

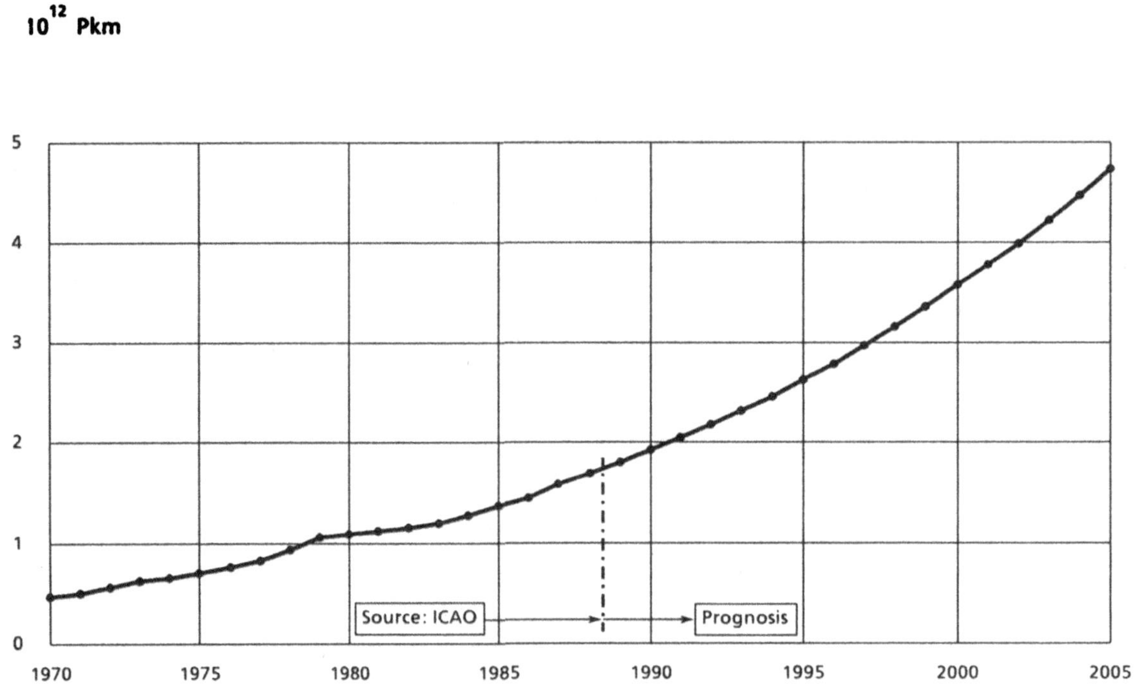

Figure 1. World scheduled air traffic (passenger tonne-kilometres transported)

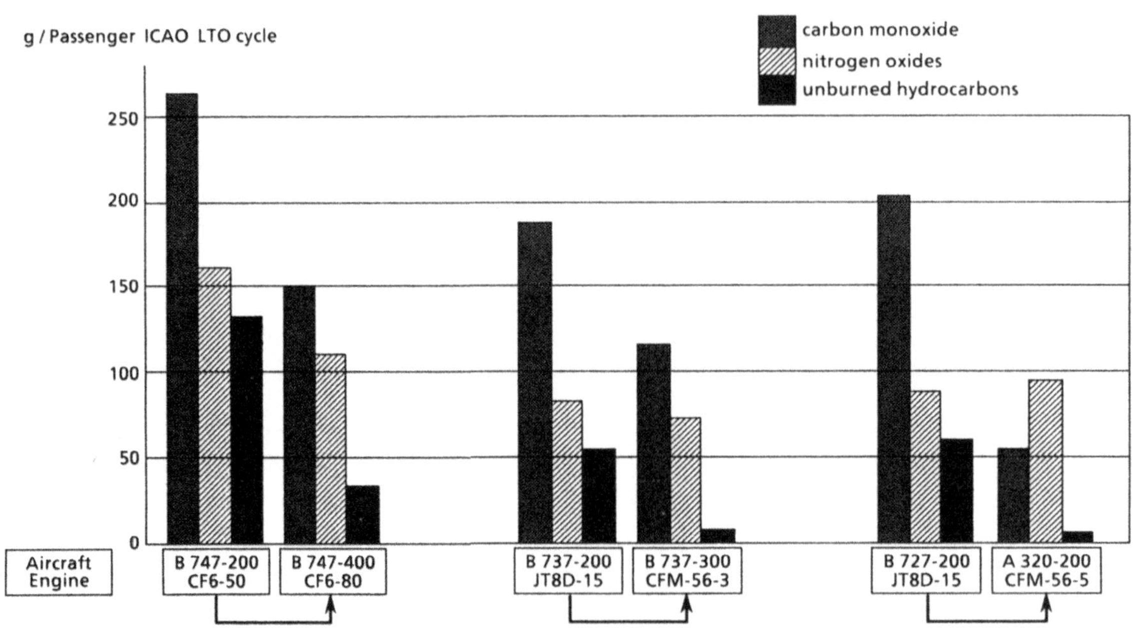

Figure 2. Reduction in emissions by fleet renewal

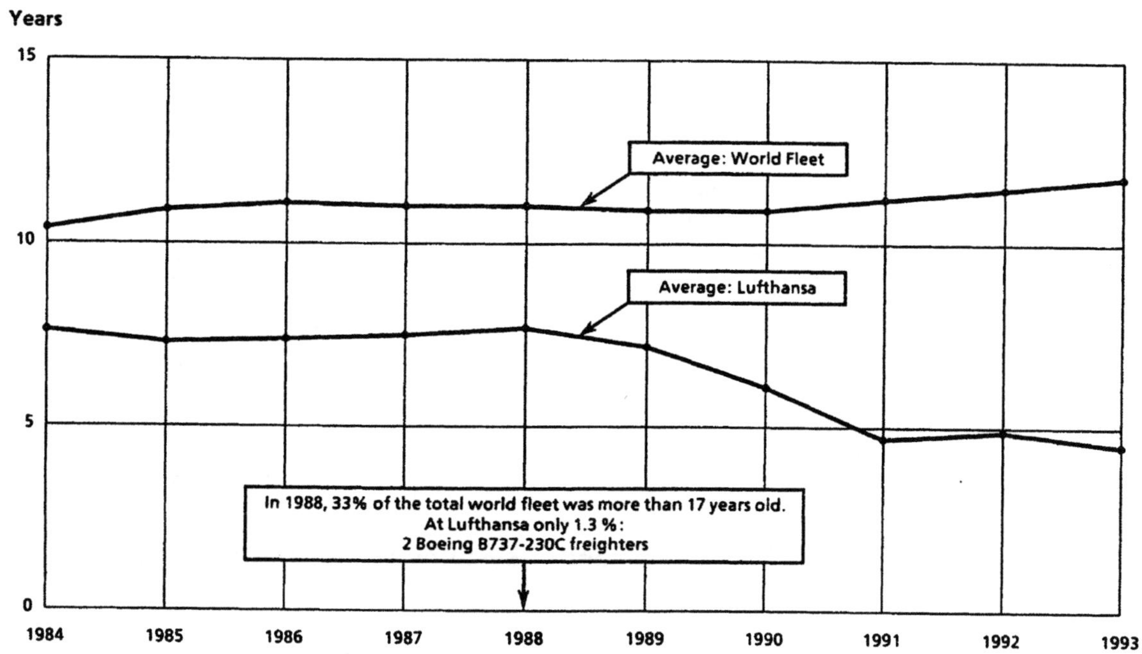

Figure 3. Age of the Lufthansa fleet in comparison to the world average

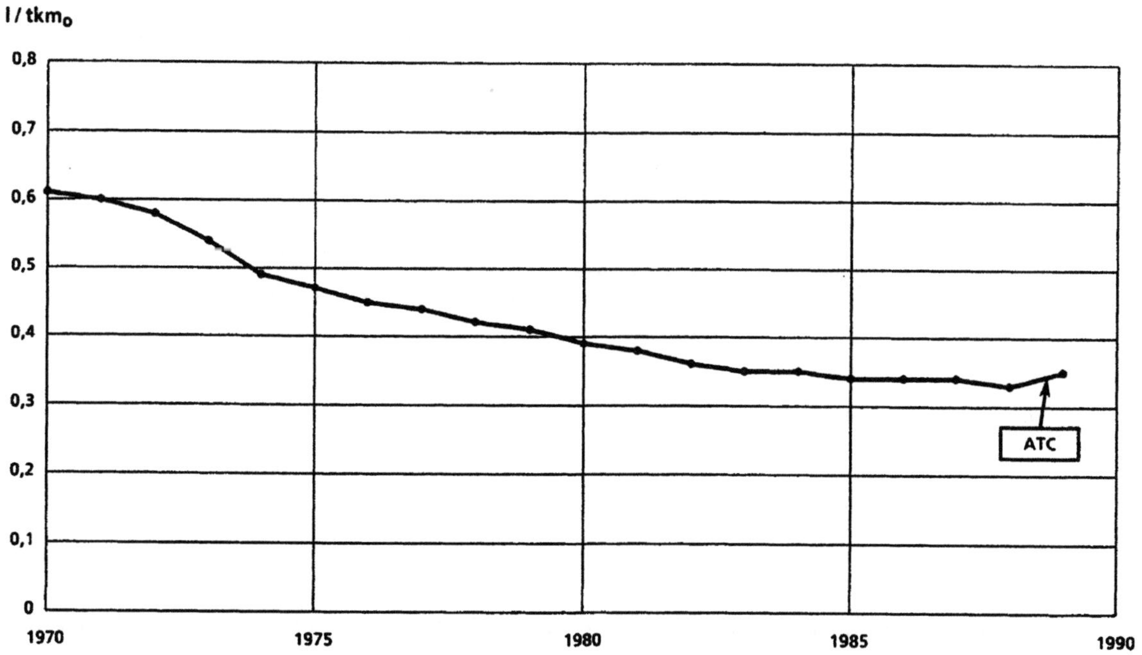

Figure 4. Lufthansa fuel consumption in litres per tonne-kilometre

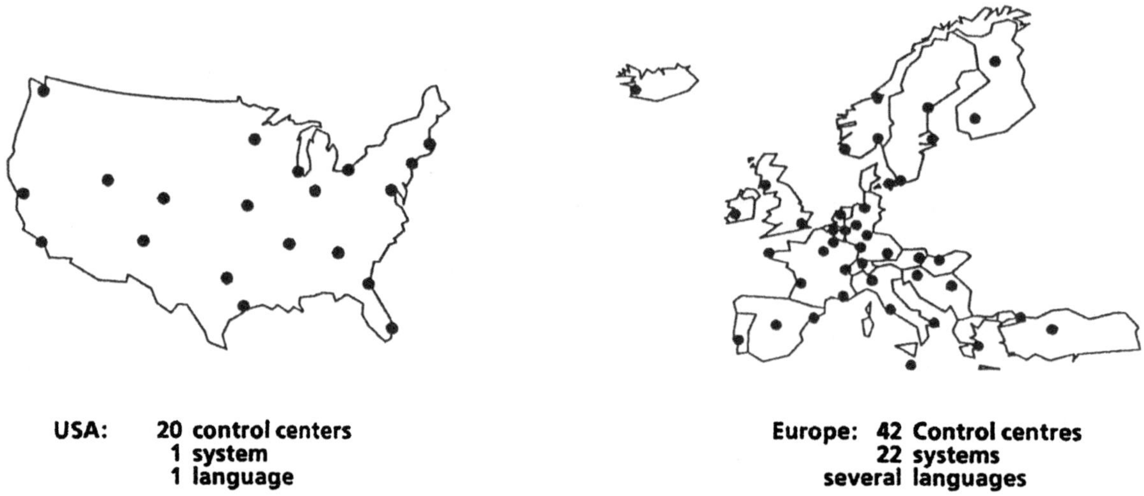

Figure 5. Comparison of air traffic control systems in the USA and in Europe

Aircraft	Engine	T.O. weight	Noise carpet
A 320-200	CFM-56-5	67,5 t	1,55 km^2
B 727-200	JT8D-15	76,2 t	14,25 km^2

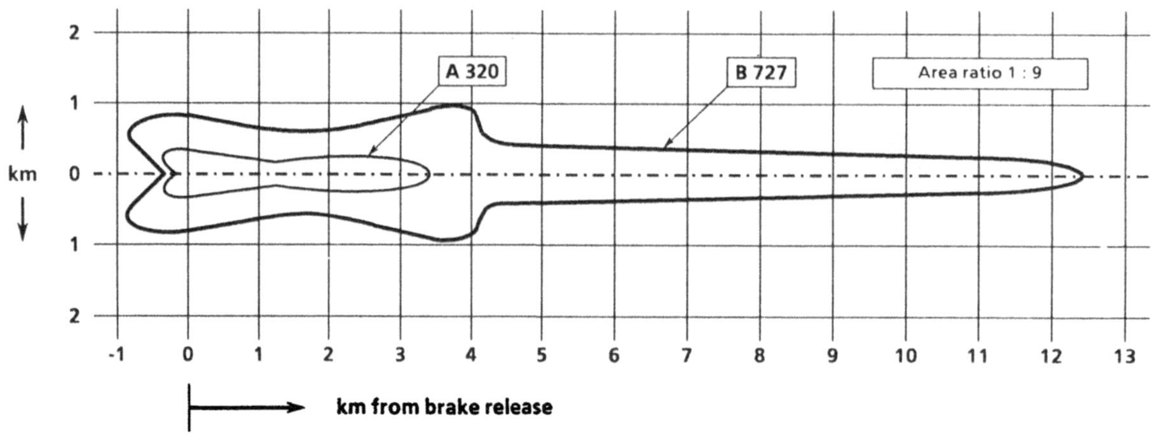

Figure 6. Noise contour at take-off of an A320 and a B727 aircraft. The contours surround the "noise carpet" with noise exceeding 85 dB(A). They apply for a flight plan of 1000 nautic miles, for take-off at sea level, and for no wind.

Aircraft Type	Engines	T.O. Thrust	Maximum T.O. Weight	Maximum Payload	Seats	Fleet Size	Number of Landings	Operation
		kN	t	t		31.12.89		since
B727	3 x JT8D-15	3 x 69	83	17	139	22	53895	
B737-200	2 x JT8D-15	2 x 69	48	12	98	38	93322	
B737-300	2 x CFM56-3B1	2 x 89	56	14	110	22	43596	
B747-200 P	4 x CF6-50E2	4 x 234	363	59	355	5	5241	
B747-200 K	4 x CF6-50E2	4 x 234	363	68	235	12	9827	
B747-200 F	4 x CF6-50E2	4 x 234	378	102	0	5	4033	
B747-400	4 x CF6-80C2B1F	4 x 258	386	60	372	3	2303	05/89
A300-600	2 x CF6-80C2A3	2 x 268	165	36	207	7	10772	
A310	2 x CF6-80A3, C2	2 x 222/238	125/153	30	199/171	17	25932	
A320	2 x CFM56-5A1	2 x 111	74	17	134	5	1690	10/89
DC10-30	3 x CF6-50C2	3 x 234	252	42	229	14	10395	
Zw'summe	-	-	-	-	-	150	261006	
Sonstiges	-	-	-	-	-	-	-	
Summe	-	-	-	-	-	-	-	

Table 1. Performance of the Lufthansa Fleet in 1989

Aircraft Type	TKO	NLF	TKT Total	TKT Passengers	TKT Cargo, Mail	PKT	Leg length (G.C.)	Fuel used
	10^6 t km	%	10^6 t km	10^6 t km	10^6 t km	10^6 P km	km	10^3 t
B727	506	54	272	241	31	2596	554	241
B737-200	550	51	282	260	22	2824	533	263
B737-300	405	54	219	170	49	1830	726	134
B747-200 P	1461	71	1031	733	298	7482	5721	413
B747-200 K	2744	73	1990	843	1147	8605	4951	686
B747-200 F	1651	74	1226	0	1226	0	5324	306
B747-400	181	67	121	80	41	820	(1609)	49
A300-600	557	64	359	243	116	2512	1708	143
A310	483	61	295	208	87	2193	717	151
A320	6	50	3	3	0	31	(241)	2
DC10-30	1572	67	1055	659	396	6727	4159	438
Zw'summe	10116	68	6853	3440	3413	35620	1133	2826
Sonstiges	1054	66	697	52	645	584	-	37
Summe	11170	68	7550	3492	4058	36204	-	2863

Table 2. Performance and fuel consumption of the Lufthansa Fleet in 1989
TKO = Tonne-kilometres, offered
TKT = Tonne-kilometres, transported
NLF = Load factor = TKT/TKO
PKT = Passenger-kilometres, transported

Aircraft Type	specific consumption	specific consumption, P = 2 Cargo, Mail	Passengers	Passengers
	g / t km	g / t km	g / t km	l / 100 P km
B727	886	470	940	10,9
B737-200	933	485	970	11,2
B737-300	612	344	689	8,0
B747-200 P	401	234	468	5,7
B747-200 K	345	242	484	5,9
B747-200 F	250	250	-	-
B747-400	(405)	(244)	(488)	(5,9)
A300-600	398	238	475	5,7
A310	512	300	600	7,1
A320	(667)	-	(667)	(8,1)
DC10-30	415	256	511	6,3
Zw'summe	412	275	549	6,6
Sonstiges	-	-	-	-
Summe	-	-	-	-

Table 3. Specific consumption of the Lufthansa Fleet in 1989

The table is computed for a specific fuel mass of 0.8 kg/litre.

Note that it is not possible to attribute consumption specifically to passengers or cargo services because all aircrafte carry cargo. In addition, payload can be limited by weight or by volume, depending on aircraft type and route. Therefore, the parameter P = 2 is assumed, i.e. Passenger tkm is calculated with double the consumption of Cargo tkm.

Emission Substance	Absolute Emissions (P = 2)			Specific Emissions (P = 2)		
	Total	Cargo, Mail	Passengers	Cargo, Mail	Passengers	Passengers
	t	t	t	g / tkm	g / tkm	g / Pkm
CO_2	8604 x 10^3	3091 x 10^3	5513 x 10^3	837	1675	162
H_2O	3380 x 10^3	1214 x 10^3	2166 x 10^3	329	658	63
C	37,2	13,4	23,8	0,00	0,01	0,00
C_xH_y	4106	1475	2631	0,40	0,80	0,08
CO	10245	3681	6564	1,00	1,99	0,19
NO_x	41445	14891	26554	4,03	8,07	0,78
SO_2	2729	980	1749	0,27	0,53	0,05

(Average for all types operated by Lufthansa)

1988 Performance:

TKT			PKT
Total	Cargo, Mail	Passengers	
10^6 t km	10^6 t km	10^6 t km	10^6 P km
6984	3692	3292	34118

TKT = Tonne-kilometres, transported
PKT = Passenger-kilometres, transported

Table 4. Specific emissions and performance of the Lufthansa Fleet in 1988

Lufthansa Flights In and Above the Tropopause

- Proportion of fuel 1988: 17 - 20 %, around 500 × 10^3 t

- For 4 actual flights on June 22, 1989, route calculations were carried out with the limitation "no tropopause contact permitted":

from	to	Additional consumption	
Düsseldorf -	Chicago	+ 4,5 t	= + 6,0%
Frankfurt -	Tokio	+ 7,7 t	= + 6,2%
Frankfurt -	Stockholm	+ 0,3 t	= + 6,7%
Frankfurt -	Anchorage	+ 8,1 t	= + 8,1%

- For flights with reduced payload, specific fuel consumption per tkm would increase even further than that.

Table 5. Lufthansa flights at and above the height of the tropopause in 1988 and additional consumption computed for four flights in 1989 under the constraint that tropopause contacts had to be avoided

HYDROGEN TECHNOLOGIES FOR FUTURE AIRCRAFT

Carl-Jochen Winter

DLR - German Aerospace Research Establishment
Energetics Research Division
Pfaffenwaldring 38-40, 7000 Stuttgart 80

ABSTRACT

Never in the history of aviation has a commercially operated hydrogen airline been set up. In the early seventies immediately after the oil crises aircraft manufacturers and national laboratories considered replacements for kerosene, which had become scarce and extremely expensive almost overnight. Now it is rather ecological threats which are the major driving forces behind a potential reorientation regarding fueling. The paper presents the technological status both in aviation and space, reflects on the appropriate international constellations and time periods required, reports on the - all-in-all - very successfully accomplished hydrogen supply state-of-the-art in space operation, and extends to considerations of infrastructure, safety, and security matters.

INTRODUCTION

In neither civilian nor military operational aviation has hydrogen had a role to date. Serious efforts have not extended beyond theoretical and experimental studies.

Now, at the end of the 20th century, there are three reasons to again seriously consider the possibilities for hydrogen in aviation:

- Conventional hydrocarbons are in limited supply. Increasing scarcity leads to higher prices, as already experienced as a temporary consequence of the oil crises of the 1970's.

- Despite its relatively small share of anthropogenic energy consumption, only a few percent, aviation contributes to the increasing pollution of the geosphere by residuals and toxic substances because this use occurs at atmospheric heights, where man is the only polluter (*Figure 1*). Despite laudable technological success in achieving a reduction in fuel consumption, the planned doubling of the world's air traffic capacity in the next 10-15 years will lead to ever more ecological conflicts.

Exhaust Emissions and Fuel Consumption

Emitters	Fuel Consumption t	%	CO t	%	VOC t	%	NO_x t	%	SO_2 t	%
Instrument Flight, Civil	1,587,020	56.7	15,540	32.5	3,900	42.6	18,700	64.7	1,590	57.9
Visual Flight, Civil	9,540	0.3	7,620	15.9	180	2.0	30	0.1	4	0.1
Military	1,205,879	43	24,663	51.6	5,076	55.4	10,162	35.2	1,152	42.0
Total Air Traffic	2,802,439	100	47,823	100	9,156	100	28,892	100	2,746	100

Civil Aviation
155 t/a Soot

Annual Emission from All Traffic Sources

Pollutant	Air Traffic						Motor Vehicles		Other Traffic		Total Traffic	
	Civil		Military		Total							
	t/a	%	t/a	%	t/a	%	t/a	%	t/a	%	t/a	%
CO	23,150	0.35	24,650	0.35	47,800	0.7	6,463,900	95.8	233,900	3.5	6,745,600	100
VOC	4,100	0.35	5,050	0.45	9,150	0.8	1,146,400	94.4	58,700	4.8	1,214,250	100
NO_2	18,750	1.1	10,150	0.6	28,900	1.7	1,450,100	86.3	201,200	12.0	1,680,200	100
SO_2	1,600	1.6	1,150	1.2	2,750	2.8	64,600	66.4	30,000	30.8	97,350	100

Source: Umweltbundesamt Berichte 6/89

Fig. 1 Air traffic in the FR Germany 1984

- Continued development plans for operational hypersonic air traffic and for the HTOL space transporter cannot be optimally realized, if at all, with hydrocarbon fuels. The gravimetric energy intensity of liquid hydrogen is three times as high as that of kerosene.

For all three reasons hydrogen could be advantageous. If produced using the non-fossil primary energies solar irradiance, hydropower or nuclear energy, hydrogen contains no carbon, no sulphur, no heavy metals; no dust, no soot is released when hydrogen is combusted. It comes from water and recombines to water. Its only potential residuals and pollutants are water vapor and nitrogen oxide, and the latter only if air serves as the oxidizer. If the global greenhouse gas reduction policy is taken up as a serious challenge for mankind in future decades, then hydrogen has a role to play in one of the few "zero CO_2" and "zero methane" technologies which absolutely does not contribute to the two major carbon-containing greenhouse gases CO_2 and CH_4, whose contributions to the greenhouse effect are currently ca. 50% and 15%, respectively.

Historical Remarks

The history of hydrogen in the aerospace industry is short: So far not a single example can be cited for a commercial LH_2 airline. Since the 1950's there have been studies and experiments, also on a 1:1 scale. For example, in 1955/56 NASA flew a B-57 "Canberra" *(Figure 2)* with one engine running on helium-pressurized LH_2 [1]. Coinciding with the 7th World Hydrogen Energy Conference in Moscow, USSR, in 1988, the Soviets experimented with a modified TU-155 *(Figure 3)* which also had one engine running on hydrogen [2]. Lockheed presented studies in the 70's for a LEAP (Liquid Hydrogen Experimental Airline Project), showing how a fleet of L-1011's could be modified to run on LH_2 *(Figure 4)* for the purpose of operating a freight airline from Pittsburgh, USA via Frankfurt, FR Germany to Riadh, Saudi Arabia [3]. Pratt and Whitney [4] experimented in the late 50's with a hydrogen engine *(Figure 5)* for a high flying reconnaissance airplane being planned at that time; however, the program quite literally never got off the ground. Of more recent date plans have resurfaced for a HALE - High Altitude Long Endurance Aircraft *(Figure 6)* [5]. Calculations show that the use of LH_2 as opposed to kerosene is

First Flight: 1956

Typical Test Parameters:

Altitude: 50,000 ft.
Speed: Mach 0.75
Duration: 17 min. with one modified J-65 engine

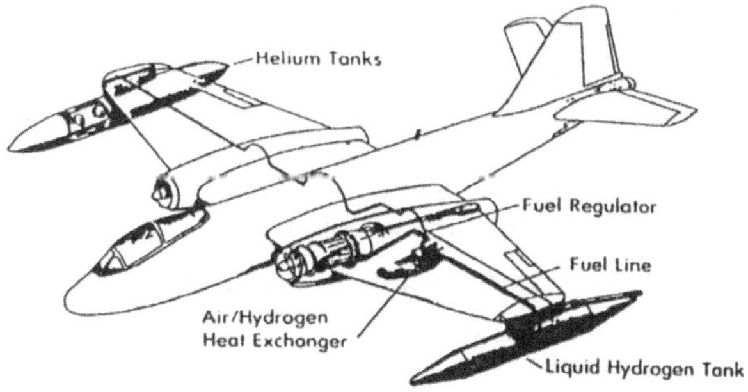

Source: NASA Lewis Research Center (1956), G. Brewer (1979)

Fig. 2 B-57 "Canberra" - first hydrogen-propelled airplane

Maiden Flight: April 15, 1988

Duration: 21 min (take off, cruise, landing)

one engine hydrogen fueled

Source: TASS (1988), Flugrevue 8/88

Fig. 3 TU-155 (modified TU-154) first hydrogen-propelled airplane in USSR

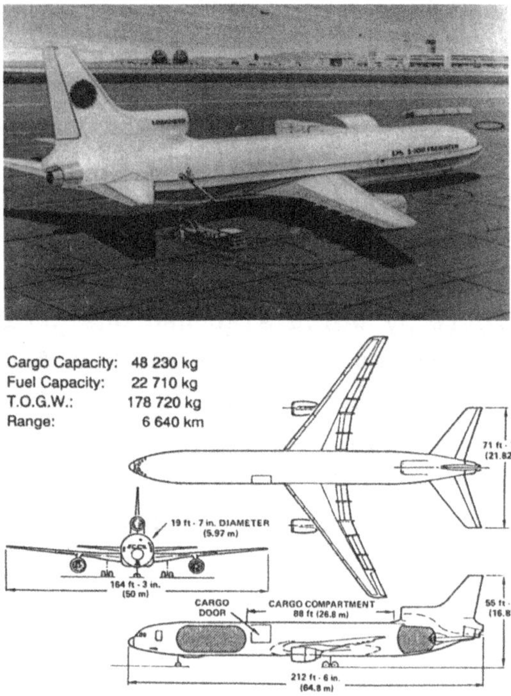

Source: Lockheed; G. Brewer (1979)

Fig. 4 LH_2-fueled L-1011 freighter (study)

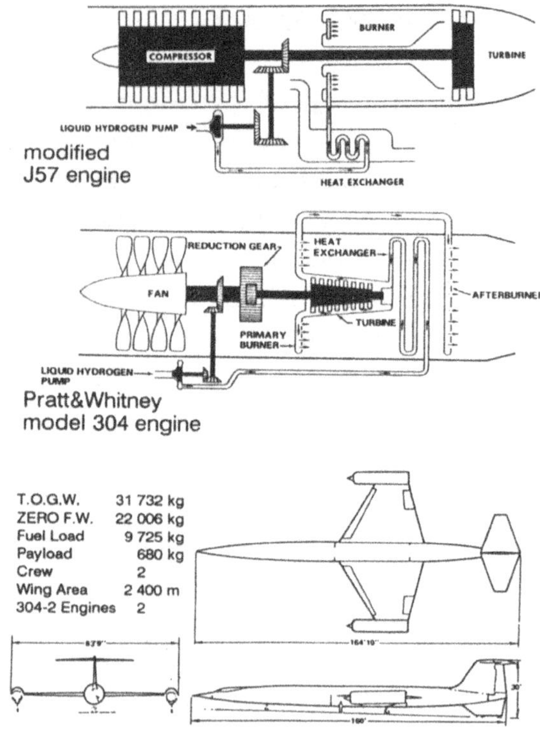

Source: R.C. Mulready, Pratt & Whitney (1979); G. Brewer, Lockheed (1979)

Fig. 5 LH_2-fueled engine prototypes for CL-400 reconnaissance aircraft study

justified by the twofold to threefold increase in mission duration. In autumn 1979 the DGLR (German Society for Aeronautics and Astronautics) and the DFVLR (German Aerospace Research Establishment) hosted in Stuttgart a four-day symposium with wide international participation on "Hydrogen in Air Transportation." The global state of the art at that time was presented [6] and an ad hoc executive group formulated and published "An International Research and Development Program for LH_2-fueled Aircraft" [7]. Its basic conclusions remain valid today, in 1990 - over 10 years later. More recently MBB has started to think about a LH_2 Airbus for a range of 1000 nmi *(Figure 7)* [8].

Subsonics, Supersonics, Hypersonics

The really new initiative which would be doomed a priori without the use of liquid hydrogen is supersonic/hypersonic flight for passenger transport in the upper stratosphere or for space flight. All countries with major space programs have such projects: With its NASP - National

Aerospace Plane - the USA is spending the most money; the FR Germany made plans in spring 1989 for a hypersonic technology program costing about 350 million DM, based on the concept of a two-stage horizontally starting and landing SÄNGER II *(Figure 8)*; Great Britain has presented plans for a HOTOL (Horizontal Take-Off and Landing) aircraft; France has proposed a new version of an advanced CONCORDE *(Figure 9)*, which recently found interest also in Britain and the US, and of an ATSF (Avion de Transport Supersonic Futur) derived from it, as well as a hypersonic vehicle, the AGV (Avion à Grande Vitesse); Japan and the USSR have their own plans. In [9] a fairly recent brief report about the NASP and its international competitors is given.

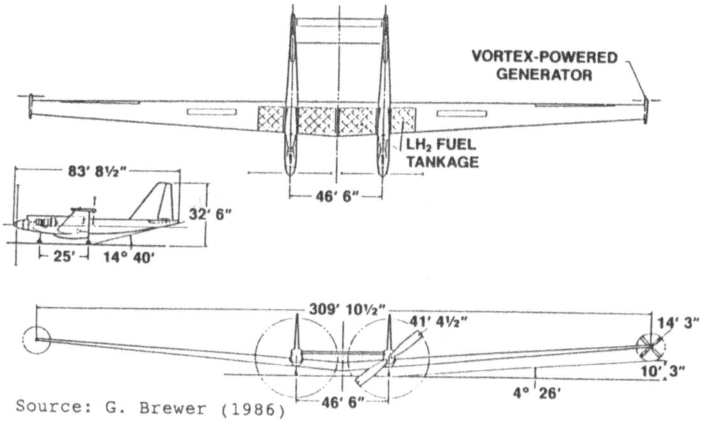

Source: G. Brewer (1986)

Fig. 6 High-Altitude, Long Endurance (HALE) Aircraft

	LH_2	JP-TS
PAYLOAD, LB.	2,185	2,185
ALTITUDE, FT.	90,000	90,000
GROSS WEIGHT, LB.	49,160	75,000
ENDURANCE, HRS.	48	18.8

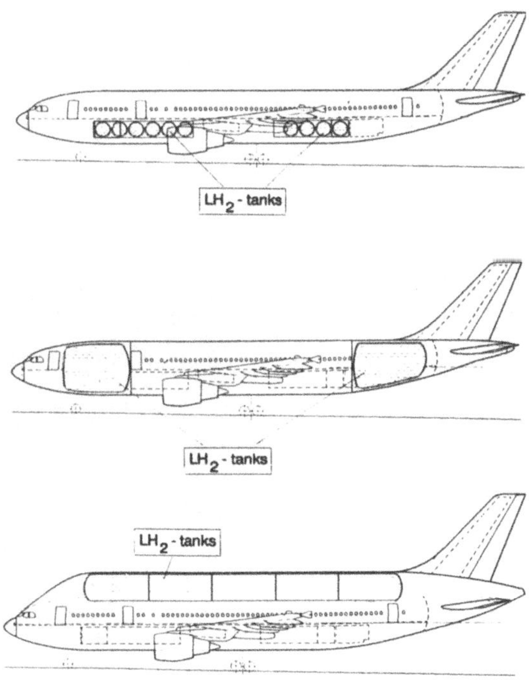

Fig. 7 Hydrogen-fueled Airbus (studies)

Source: MBB-UT (1989)

Aerospace and hydrogen

If in the aviation field it is only plans and proposals, or at best prototypes, which can be referred to, there is more to say about the space technology area: Space transportation is no longer conceivable without operational, high energy LH_2 /LOX engines. In fact, this is the only industrial realm worldwide in which the *energetic* use of hydrogen is absolutely necessary. (All the other realms use hydrogen non-energetically, or only quasi-energetically, whether for the production of fertilizers or of methanol, for the petrochemical, fat hardening or electronics industries, for the cooling of electric generators, etc.).

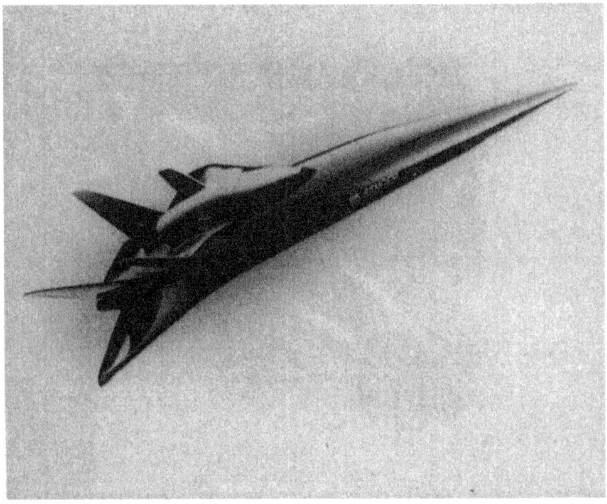

Main Data (1988)

Total Launch Mass	**340 Mg**
Vehicle Length (fuselage)	81.3 m
Wing Span	41.4 m
Hypersonic L/D	4.8-5.3
First Stage (EHTV)	**259 Mg**
Nominal Net Mass	149 Mg
Maximum Propellant Mass (LH)	100 Mg
Engine Thrust (ground)	5*300 kN
Max. Speed (on separation)	Mach 6.8
Cruise Range	2*3700 km

Second Stage Horus	
Nominal Net Mass	22.2 Mg
Propellant Mass (LOx/LH_2)	65.5 Mg
Engine Thrust (vac)	1200 kN
Payload (2-4)	3.3 Mg
Second Stage Cargus	
Nominal Net Mass	7 Mg
Propellant Mass (LOx/LH)	58 Mg
Engine Thrust (vac)	1050 kN
Payload (2-4)	5-15 Mg

Source: MBB-UT (1989)

Fig. 8 Sänger II system

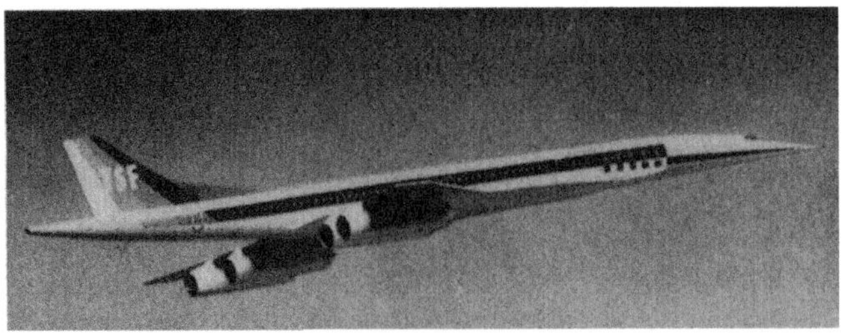

(in discussion)

Avion de Transport Supersonique Futur
Passengers: 200
Range: 12 000 km
Speed: Mach 2-2.5

Source: Aérospatiale (1988)

Fig. 9 Concorde successor: ATSF

On the other hand, all major space nations either have engines running on the high-energy fuel combination of LH_2/LOX already in their rocket programs or they are developing them *(Figure 10)*: The USA for the space shuttle; Europe for the ARIANE V; the USSR for its ENERGIJA, which just a few years ago was used for the first time and for whose transport services there is talk of as low as 1,650 US$/kg LEO, and Japan for its H-II - to give only a few examples. All the knowledge gained from space hydrogen technologies can benefit aviation - and every bit of it will be needed, for example in thermodynamics and reaction kinetics, for cryotechnology, and for developing a hydrogen infrastructure as well as procedures for handling hydrogen, not to mention hydrogen safety and security. Only one area, should it develop, belongs exclusively in the realm of aviation: The use of the large, on-board heat sink for cryolaminarization to reduce drag and cool engine components. However, once this technique will have been used and perfected in aviation, it will be sure to find a role in space technology, for example if reusable, winged vehicles become the rule.

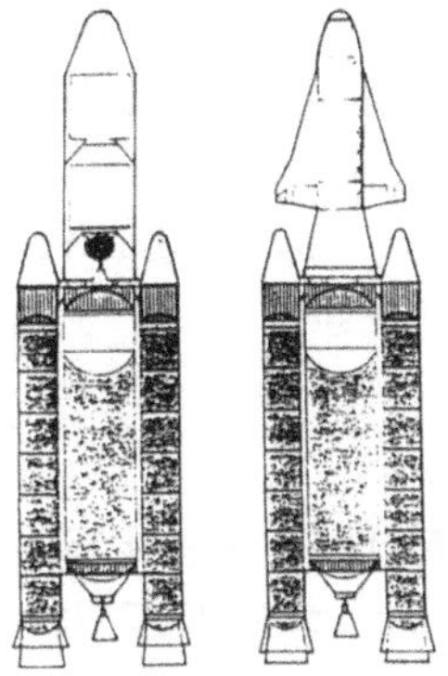

ARIANE 5 with H155

Name	Country	Engines (number)	Thrust (kN)	1.Flight	Status
Centaur	USA	RL-10 (2)	133	1962	operational
S-4	USA	RL-10 (6)	400	1964	1964 - 1965
S-4B	USA	J-2 (1)	1023	1966	1966 - 1975
S-2	USA	J-2 (5)	5120	1967	1967 - 1973
H-10	ESA	HM-7B (1)	62	1979	operational
H-8	China	YF-73 (1)	49	1984	operational
LE-5	Japan	LE-5 (1)	117	1986	operational
Shuttle	USA	SSME (3)	5005	1981	operational
Energija	USSR	(3)	5340	1987	operational
H-18	China	YF-75 (1)	78		planned
LE-7	Japan	LE-7 (1)	910		planned
H155	ESA	HM60 (1)	1050		planned
H-70	Germany	ATC-700 (1)	700		planned

Source: Space World (1/88)

Fig. 10 Rocket stages powered by liquid hydrogen

Time Periods and International Aspects

Aircraft development from project beginning to first test flight - especially if the developments break new ground - takes 10 years, and often longer. If the plane goes into production, it will be in operation for 20 to 25 years. Applying these figures to the first-ever hydrogen fueled fleet when it becomes a reality, regardless of how small the numbers, this means that if a start is made in the early 90's, it would be very optimistic to expect the first test flight before the end of the century, if that early. Assuming that the development work produced positive results, production could begin in the first decade of the new century; the first planes could be in the air between 2010 and 2020, and the last ones would be grounded in 2030-40 - a time period which would include two to three generations of engineers and, for democratic countries, innumerable election cycles. National solo attempts are unthinkable. International, even global thinking is called for in the development, construction, financing and operation of the entire system.

It is hard to imagine that a hydrogen-fueled aeronautical or space transportation system of huge costs and long time scales can arise, or even be operated, in international competition. The required technology can be developed in competition; this will help the best ideas to surface. But to establish such a system, to equip it with infrastructures and to operate it can probably only be done once, or if international politics does not allow for this, then twice at best (West and East), and in international cooperation. - In [10] a hydrogen aircraft penetration study is presented which starts the market penetration in 2000 and continues it as far as 2075; special emphasis is paid to the amount of CO_2 not emitted by switching at an accelerated pace from kerosene to liquid hydrogen.

However, it should be noted that the EUROMART Study [11], published in April 1988 and prepared jointly on behalf of the EEC by all important European aircraft producers, mentions hydrogen aircraft and second generation supersonic transport vehicles as well as hypersonic/trans-atmospheric vehicles only in passing. This suggests that the participating European aircraft industries have set their standards for EUROpean Cooperative Measures for Aeronautical Research and Technology with an eye to the near, rather than the distant, future.

Technologies, Economics Ecologies

In addition to these considerations, it should be noted that in the case of the first hydrogen airplane worldwide (whether subsonic, supersonic or hypersonic) we are not at all dealing with a new model in an otherwise familiar series, but with a vehicle that in many respects poses challenges for which there are no, or only partially applicable, lessons to be learned from precursors - perhaps from missiles, for example. Here one could mention such challenges as:

- Combination engines which simultaneously meet the specifications of air breathing subsonic or supersonic operation, of high altitude supersonic operation, as well as of rocket engines in the hypersonic regime.

- On the one hand cryotemperature tanks (20°K), and on the other hand lightweight airframe structures which can withstand high temperatures ($\leq$ 2000°C) and which must make payloads possible of a few percent of the total weight.

- Designing materials and structures which can withstand an aggressive hydrogen atmosphere.

- Since hydrogen-fueled aircraft will presumably be bulky because of the larger tank volume, and on the other hand since there are large heat sinks in the liquid hydrogen tanks, all developments which can reduce drag are especially important for LH_2 aircraft:

  - Laminar flow control by sophisticated aerodynamic design, active boundary-layer control through skin suction or surface cooling by the on-board cryogenic liquid

  - Riblets, etc.

- A completely new and safe-to-operate hydrogen infrastructure on the ground, developed and maintained parallel to the hydrocarbon infrastructure which will continue to exist.

Two particular conditions may be difficult (if not impossible) to fulfill, despite a clear political will and the availability of the necessary engineering and financial expertise:

(1) Having available the required amounts of liquid hydrogen and liquid oxygen at competitive prices: This condition will probably need to be set primarily by the hydrogen *aviation* industry, which will not suddenly cease to use inexpensive kerosene or even synthetic

hydrocarbon jet fuel, e.g. made from coal (Synjet), the moment the first hydrogen airplane appears in the skies. Two completely independent infrastructures, an already-established hydrocarbon infrastructure and a yet-to-be established hydrogen infrastructure, will need to exist parallel to each other and to be maintained for many decades! It will be all the harder for hydrogen to find a place in subsonic civil aviation the more the development of kerosene-fueled aircraft with low specific fuel consumption is successful *(Figure 11)*. From 1960 to 1989 alone, specific consumption has been reduced by 40%, and an additional 20% can reasonably be expected. It is trivial to say that the kilogram fuel not needed per kNewton thrust and hour does not pollute the environment. And, according to laboratory-scale experiments, the potentially expected NO_x reduction potential is still impressively high: Minus 30% with lean combustion, minus 70% with rich/lean combustion management with intermediate quenching, and even minus 85% with lean combustion and premixing/pre-vaporizing of the fuel/oxidizer mixture [12].

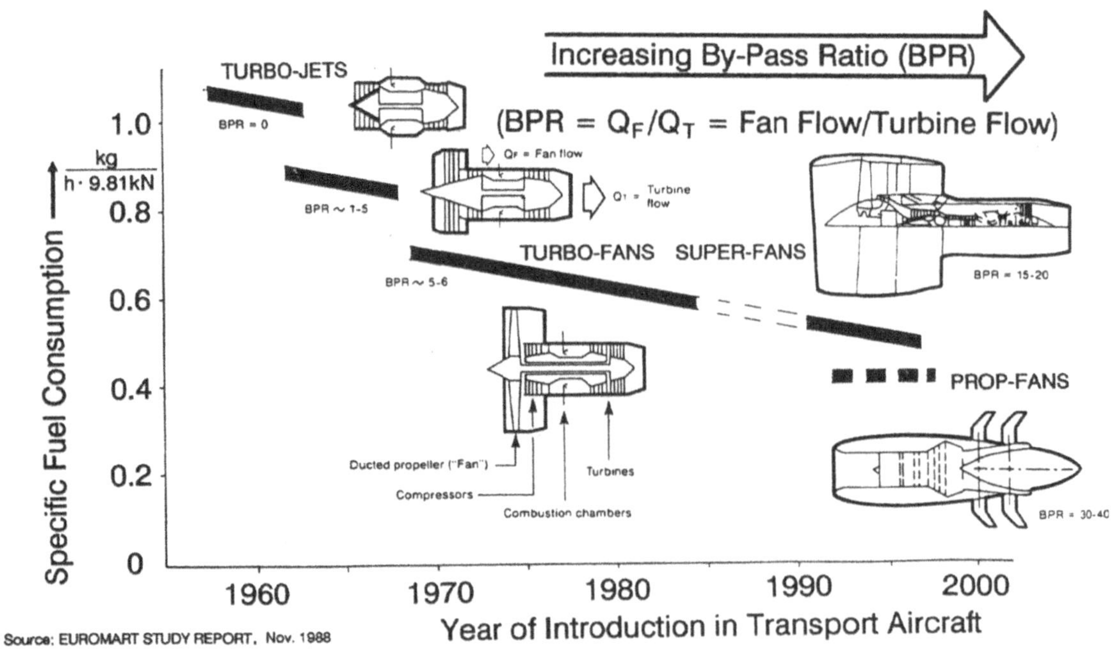

Fig. 11 Progress on propulsion

(2) Guaranteeing the ecological acceptability also of high-speed hydrogen airplanes, which for economic reasons should regularly operate in the upper troposphere, in or above the tropopause, or even beyond: At these heights man is almost the only polluter, and not only nitric oxides, but also water vapor may become environmentally harmful there. As water vapor freezes, the ice crystals formed collect in cirrus-like clouds which expand slowly and persist a long time. [13,14,15]. If it is confirmed that high-flying hydrogen aircraft contribute to an intolerable extent to the anthropogenic greenhouse effect, there would still be the possibility of flying at lower, albeit uneconomical, levels.

The space industry does not have this option. All its vehicles, including hydrogen-fueled space vehicles, inevitably produce water vapor in their ascent (and descent) through the troposphere and stratosphere, although the generation of nitric oxide is being prevented by using liquid oxygen as the oxidizer. This may mean, for example, restricting for ecological reasons the frequency of trips to service permanently manned space stations.

Hydrogen safety and security

Whenever large amounts of liquid hydrogen have to be handled, then questions arise about safety, and about security from sabotage or terrorist activities. It is true that when compared with kerosene, hydrogen has certain undesirable specific physical, chemical, or reaction kinetic characteristics. These must, however, be weighed against hydrogen's definitely desirable characteristics (*Figure 12*). Indeed, the flammability of a flammable hydrogen-oxidizer mixture is high. For instance, hydrogen ignites quickly enough to operate a scramjet with air passing through it at Mach numbers of 3, 4 or greater. The ignition range is large, and there is a high tendency to deflagrate in closed areas. But on the other hand, flame and diffusion velocities are high in air, so that burning time is short and ground level horizontal spreading of the fire is minimal. And most important (*Figure 13*): Since toxic substances are not involved, there is no chance of poisoning or suffocation as a consequence of the fire. - The security problem is one that is common to *all* centralized technical systems, and in our case, a hydrogen aeronautical/space system, it is of worldwide dimensions, to be sure. One will have to think of solutions similar to those already worked out at present

	Hydrogen	Methane	Jet A	JP-4
Nominal Composition	H_2	CH_4	$CH_{1.93}$	$CH_{1.93}$
Molecular Weight	2.016	16.04	~ 168	~ 132
Heat of Combustion (low), kJ/g	120	50.0	42.8	42.8
Liquid Density, g/cm³ at 283 K	0.071*	0.423*	~ 0.811	~ 0.774
Boiling Point, K (at 1 Atmosphere)	20.27	112	440 to 539	333 to 519
Freezing Point, K	14.4	91	233	215
Specific Heat*, J/g K	9.69	3.50	1.98	2.04
Heat of Vaporization, J/g (at 1 Atmosphere)	446	510	360	344
Vapor Pressure, kPa	-	-	-	18
Diffusion Vel. in NTP Air, cm/s	≦ 2.00	≦ 0.51	< 0.17	< 0.17
Buoyant Vel. in NTP Air, m/s	1.2 to 9	0.8 to 6	Non Buoyant	Non Buoyant
Flammability Limits in Air, Vol.%	4.0 to 75.0	5.3 to 15.0	0.6 to 4.7	0.8 to 5.8
Vaporization Rate W/O Burning, cm/min (From Liquid Pool)	2.5 to 5.0	0.05 to 0.5	N/A	0.005 to 0.02
Min. Ignition Energy in Air, mJ	0.02	0.29	0.25	0.25
Autoignition Temp., K	858	813	> 500	> 500
Rate of Liquid Lowering of Burning Pool, cm/min	3 to 6.6	0.3 to 1.2	0.17	0.20
Burning Vel. in NTP Air, cm/s	265 to 325	37 to 45	18	381
Flame Temp. in Air (Stoichiometric), K	2318	2148	2200	2200
Thermal Energy Radiated to Surroundings %	17 to 25	23 to 33	30 to 42	30 to 42
Detonability Limits in Air, Vol.%	18.3 to 59.0	6.3 to 13.5	N/A	1.1 to 3.3

* At Normal Boiling Point; NTP = Normal Temperature and Pressure

Source: G. Brewer et al. (1981), R. Mc Carty et al. (1981), ASTMD-1655-80a Jet A, Mil-T-5624L, EXXON (1973), J. Kuchta (1973), H. Barnett (1956)

Fig. 12 Selected properties of fuels

Liquid hydrogen poses substantially less hazard to crew and passengers, as well as to persons and property in adjacent areas, in event of a crash.

- **In an otherwise survivable crash, LH_2 tanks are less apt to be ruptured, thus minimizing spill potential.**
- **If fuel is spilled, LH_2 vaporizes, becomes buoyant almost immediately, and dissipates in the atmosphere. Time and area exposed to hazardous concentrations are both minimized.**
- **If spilled fuel is ignited, duration of H_2 fire will be extremely short and heat-affected area much smaller than fires with hydrocarbon fuels.**
- **Probability of fire with spilled fuel in a crash is about equal with any aircraft fuel.**

Conclusions of studies conducted for NASA-Lewis Research Center by Lockheed-California Co. and A.D. Little, Inc.

Source: G.D. Brewer (1986)

Fig. 13 Crash fire hazard

for large international airports or space centers. In Europe, for example, a detailed study was made of hydrogen operation at Zurich international airport in Switzerland [16]. Certainly special risks have to be taken into account when planning for the production, transport and storage of huge amounts of hydrogen on land, sea and in the air. But there is one consolation: Toxic substances, as mentioned above, and radioactivity (as it occurs in nuclear energy systems of comparable energetic potential) *cannot* arise per se. So there is *no need* to fear the eventual occurrence of any health problems of unknown duration related to hydrogen. The relevant U.S.-American literature is full of quotations from extensive studies and experiments on the issue of safety [5,6,17,18]; attention is specifically called to them at this point.

Hydrogen supply

Establishing a role for liquid hydrogen (or liquid oxygen) in the aerospace industry does not simply involve a switch from one fuel to another. The fluid part of the global hydrocarbon energy economy, of which the aerospace industry is a small, albeit high-tech, part, has existed basically unchanged for 100 years. There is a whole realm of knowledge and experience, infrastructures and investments, norms and procedural regulations which have come to be taken for granted and against which a new global hydrogen energy economy will inevitably be measured. That part of the world's total end energy consumption which is attributable to energetic use in the aerospace industry, 2.6% in 1986, is quite small, but the proportion of end energy consumption from hydrocarbons, 90% of the total, or $7.5 \cdot 10^9$ tce (1988), as compared to barely $500 \cdot 10^9$ Nm^3 for hydrogen per year (1986) ≙ $185 \cdot 10^6$ tce/a, makes a clear point: $185 \cdot 10^6$ tce/a are equivalent to just about two thirds of the end energy consumption of the FR Germany *alone*! A completely new hydrogen infrastructure would have to be implemented for the aviation industry, with support centers around the world similar to today's space centers in Cape Canaveral, USA or Kourou, French Guiana, or to the rocket test centers in Vernon, France or in Lampoldshausen, FR Germany (*Figure 14*), the last two being at present the largest European consumers of liquid hydrogen, using several $100 m^3$ LH_2 per week. Just for comparison, Frankfurt, the second largest international airport in Europe, needs an average Jet-A kerosene supply of 5,000 t/d. Accordingly, approximately 1,600-1,700 t LH_2/d would be needed if all incoming and outgoing aircraft would have to be fueled with LH_2. - *Figure 15* shows the test run of the 1000 kN MBB combustion chamber for the Ariane V engine in Lampoldshausen.

Fig. 14 Aerial view of DLR facilities, Lampoldshausen, ARIANE V project

Source: MBB (1989)

Fig. 15 DLR Lampoldshausen, ARIANE V project - test run of 100 kN combustion chamber on P3.2 test stand

All these users receive the required quantities of hydrogen in liquid form from far distant liquefaction plants which are usually part of the technical gases industry. Liquefaction is by the well-known Joule Thompson/Claude Process at about one third of the Carnot efficiency *(Figure 16)*. Magneto-caloric liquefaction, which makes use of the temperature drop which occurs in the case of certain rare earths when they are alternately magnetized/demagnetized, promises an increase in the efficiency of liquefaction, but is worldwide still at the laboratory stage. A special research and development challenge is the use of hydrogen slush, a mixture of liquid hydrogen and hydrogen ice. A decrease in the vapor pressure down to the triple point gives rise to hydrogen ice, which detaches itself from the tank walls and sinks to the floor if the pressure is again slightly raised above the triple point. Hydrogen slush is 15-20 percent denser than liquid hydrogen and its ability to absorb heat, and therefore to cool, is higher. Its combustion characteristics are good. Used in a hypersonic vehicle, hydrogen slush would mean a reduction in aircraft size [19].

All possible surface transportation methods are put to use. Cape Canaveral receives its LH_2- and LOX supplies from New Orleans, previously

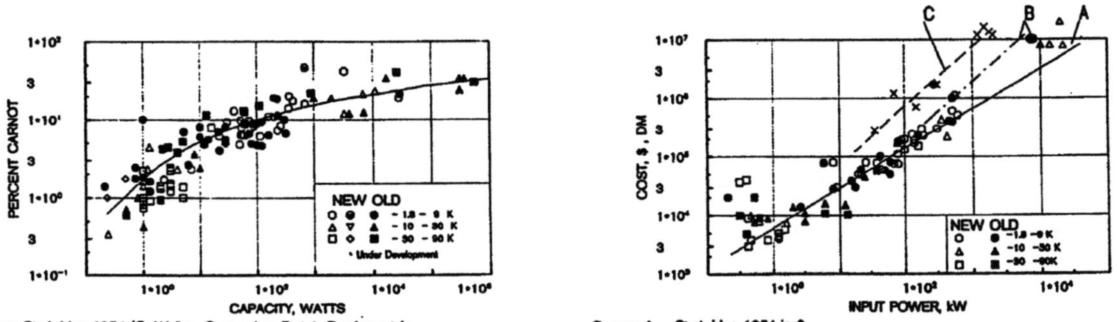

Source: Strobridge 1974 (G. Walker, Cryocoolers Part 1: Fundamentals, Plenum Press, New York)

Source: A Strobridge 1974 in \$
B J.F. Parmer, M.W. Liggett, Magnet cooling economics, IEEE Trans. o. Magn.
C KfK 1987 (x LHe-refrigerators, + LN$_2$-refrigerators), cost in DM

Fig. 16 Efficiencies and cost of cryogenic cooling systems

by ship via the Caribbean or by rail, today by road. The US space program was and is the largest user of LH_2: In the last ten years, it occupied 90% of the total LH_2 market in the US, e.g., 800 · 10^6l LH_2 flew in the space business of NASA or DOD. The shuttle alone needs 1.5 · 10^6l per fueling, which is the equivalent content of approximately 45 tanker trucks. Several 30-60 t LH_2/d facilities are operating in the US, delivering LH_2 at plant gate prices of ca. 0.25 US\$/l; the largest liquefaction facilities in Europe are in France and Holland. Kourou's supplies come by sea; Vernon's and Lampoldshausen's by road from locations within Europe. Any local liquefaction facilities usually exist only for the purpose of reliquefying boil-off losses.

It is doubtful that this logistic system can be maintained when the aviation, and especially the space industry, have to satisfactorily operate a future worldwide hydrogen enterprise with frequent take-offs and landings. The reason: Almost all commercial hydrogen supplies today come from fossil energy resources; only a small percentage is produced electrolytically, and then usually only wherever (surplus) low-cost electricity is available, for example from large hydroelectric dams (Aswan or Norway). But "fossil hydrogen" hardly meets the ecological expectations of a clean energy carrier in a post fossil fuel era. These expectations can only be met by electrolytic hydrogen, produced with the help of electricity from "solar" hydropower plants, solar power plants, or, if politically possible and safe enough, nuclear power plants. Since the large hydropower reserves on earth are highly concentrated and discretely located in places like Greenland, Canada, South America, Africa and Asia, and since large solar power stations will preferably be located in the equatorial belt of ±40°N/S - in other word, precisely *not* where heavy air traffic connects the industrialized countries of the north - transportation of this "solar hydrogen" over extremely long distances to the aerospace centers of the northern hemisphere is mandatory. This is not the case for nuclear power plants. In principle they can be located next to any airport, any space center, assuming, as stated above, that this is defensible from safety considerations and politically possible.

An unflattering, but perhaps all the more compelling argument for the promotion of the combination "hydrogen-aerospace" could become the most important one of all: Aerospace and commercial aviation as pilot branches operated exclusively by experts, not by laypeople, in the transition to a future world hydrogen energy economy. There are many indicators for the large role which hydrogen will play in global energy supply [20,21]. But the transition from a well established global hydrocarbon energy economy to a potential global hydrogen energy economy needs trailblazers which are as inherent as possible. Such a trailblazer is the use of a storable, chemical energy carrier in transportation, one which approaches the advantages of hydrocarbons, such as easy storage, and avoids its disadvantages, such as ecological incompatibility: Hydrogen. Increasing its visibility as an energy carrier for people who will be dealing with it in future would have a synergy effect. This implies its use in hundreds of millions of automobiles, and in aerospace, where it will by then have been used for decades almost without accidents. In aviation, it will need to be used operationally for several decades, and professionally by experts in its production, storage, transport, liquefaction, combustion and - not least - safe handling, By demonstrating its practicability in an area that is spectacular as far as public awareness, the establishment of a hydrogen energy economy would receive an invaluable boost on the one hand, and by its establishment in all facets of a necessary global hydrogen infrastructure, on the other hand, the aerospace industry would finally obtain the foundation, in the literal sense, needed for its success. In fact, the world is already on the way to increasing the hydrogen/hydrocarbon ratio: The H/C ratio of coal is 1, of mineral oil 2 and of natural gas, which consists primarily of methane, 4. In the transition to a hydrogen energy economy which incorporates ever more hydrogen elements, this ratio approaches infinity. Of course this does not mean that developments *have to* continue this way; but it is very probable that they will: And hydrogen subsonic aviation, hydrogen supersonic aviation, as well as, potentially, hydrogen hypersonic space flight will continue the trend [22].

CONCLUDING REMARKS

Hydrogen is no stranger. Space transportation and hydrogen are today inseparable. And there are no unsurmountable difficulties with using hydrogen in aviation. At high supersonic and hypersonic levels hydrogen is superior to its hydrocarbon competition. Safety concerns can be dealt with. In most hazard scenarios, hydrogen is safer than the hydrocarbons in current use.

If hydrogen has to be produced from fossil energy resources, which are in addition non-renewable, it cannot be considered a clean energy carrier. Clean from one end of the conversion chain to the other, from primary energy via secondary energy to end energy is only solar hydrogen, produced electrolytically with the help of solar power plants or hydropower plants, or, in the equatorial belt, OTEC - Ocean Thermal Energy Conversion plants.

The *use* of hydrogen is ecologically clean if accompanied by LOX so that nitric oxide formation can be prevented, and if in aviation hydrogen-fueled planes fly far enough under the tropopause so that, when man-made water vapor is added to the naturally present water vapor, it does not contribute significantly to the greenhouse effect. However, operational space transportation means that for every take-off (landing) the entire troposphere-tropopause-stratosphere system *has to* be flown through once (twice). Space flight is the only anthropogenic source of pollution in the upper levels, since even exhaust water vapor is viewed to become an effective greenhouse gas there.

Just as it took decades to set up a space infrastructure, the establishment of an aviation-hydrogen infrastructure will require many decades, be the work of 2-3 generations of engineers, and continue over a large number of political legislative periods in democratic countries. Continuity is needed. And international cooperation: Because aviation and space flight are truly worldwide and, one suspects, not to be financed - any more - by even the richest national economies. Space flight has so far cost hundreds of billions of U.S. dollars; hydrogen aviation and space flight will cost hundreds of billions of U.S. dollars more before they are proved operationally successful.

The introduction of a general world hydrogen energy economy, of which aviation and space flight would be components, would have a positive effect. Other human paradigms which might respectively be considered more urgent, such as development of the South, democratization and economic as well as ecological revitalization of the socialist countries, or restructuring the global economic system into one of environmental compatibility and qualitative growth, could have a counteracting effect. Because the resources which mankind is able, and willing, to muster will not be sufficient for everything.

The scientific and technical challenges are extremely appealing, whether or not the ultimate goal can be achieved. But the ability to shape a new system has to be coupled with the desire to do so. That something is technically possible is not a guarantee for success. The entire system will have to be optimized, and that involves many *non-technical* factors. We are faced with the task of making interrelated technical and non-technical choices. The "quest for the best" [23] is necessarily a quest for the *technically* best, but satisfactory is only the quest for the *technologically* best, which has to take into account the economy and ecology, safety and technology assessments - and many additional factors!

BIBLIOGRAPHY

Cited in Text:

[1] Conrad, E. William, "Turbine Engine Altitude Chamber and Flight Testing with Liquid Hydrogen" in [6].

[2] Hillebrand, Helga L., "Dampf-Flugzeug", Flug-Revue, 8/1988, p. 81.

[3] Brewer, G.D., "The prospects for liquid hydrogen fueled aircraft", Int. J. Hydrogen Energy, Vol. 7, No. 1, 1982, pp. 21-41 and "A plan for active development of LH_2 for use in aircraft", Int. J. Hydrogen Energy, Vol. 4, 1979, pp. 169-177.

[4] Brewer, G.D., "Hydrogen-fueled aircraft" in Hydrogen: Its Technology and Implications, Vol. 4, CRC Press, 1979.

[5] Brewer, G.D., Considerations Regarding Use of LH_2 in Aircraft, Hydrogen Technology Conference, 4/5 June 1986, NASA Lewis RC.

[6] DGLR/DFVLR International Symposium, Hydrogen in Air Transportation, 1979, and Hydrogen in Air Transportation - Supplement, 1979.

[7] Hydrogen in air transportation ad hoc executive group, An international research and development program for LH_2-fueled aircraft, July 1980.

[8] "Pilotprojekt Airbus mit Wasserstoffantrieb", Messerschmitt-Bölkow-Blohm, TE155-060/89; March 1989.

[9] Moteff, J.D. and Radzanowski, D.P., NASP National Aerospace Plane, The Library of Congress IB 89128, October 20, 1989.

[10] Victor, D.G., Liquid Hydrogen Aircraft and the Greenhouse Effect, International Journal of Hydrogen Energy, Vol. 15, No. 5, 1990.

[11] EUROMART Study Report, Specialised Printing Services Ltd., Essex, U.K., Nov. 1988.

[12] VDI-Nachrichten 18 May 1990.

[13] Pfeiffer, M. and Fischer, M. (Hg.), Unheil über unseren Köpfen? - Flugverkehr auf dem Prüfstand von Ökologie und Sozialverträglichkeit, Quell Verlag, 1989.

[14] Held, M. (Hg.) Ökologische Folgen des Flugverkehrs, in Tutzinger Materialie Nr. 50/1988.

[15] DLR-Vermerk, Schumann, U, "Angaben zum Luftverkehr und seine Umweltwirkung"

[16] Alder, H.P., Hydrogen in air transportation, feasibility study for Zurich airport, Switzerland, EIR-Bericht Nr. 600, September 1986.

[17] Hydrogen Safety Committee, National Research Council Canada, Safety Guide for Hydrogen, 1987.

[18] Brewer, et al., "Assessment of Crash Fire Hazard of LH_2-Fueled Aircraft" NASA CR-165525, Dec. 1981.

[19] Smith, E.M., "Slush hydrogen for aerospace applications", Int. J. Hydrogen Energy, Vol. 14, No. 3, 1989 pp. 201-213.

[20] Peschka, W., Flüssiger Wasserstoff als Energieträger, Springer, 1984

[21] Winter, C.-J., Nitsch, J. (Eds.), Hydrogen as an Energy Carrier, Springer, 1988

[22] Marchetti, C., "The future of hydrogen - an analysis at world level with a special look at air transport", Int. J. Hydrogen Energy, Vol. 12, No. 2, 1987, pp 61-70.

[23] Zimmerli, W. Ch., "The Quest for the Best. On the Responsibility of Scientists and Engineers" in Bergmann, H.W. (Ed.), Optimization: Methods and Applications, Possibilities and Limitations, Springer, 1989, pp. 1-21.

For Further Reading

1. Martin, James A., "Comparing Hydrogen and Hydrocarbon Booster Fuels", J. Spacecraft, Vol. 25, No. 1, pp. 92-94.

2. NASA Contractor Report 174840, Jet fuel property changes and their effect on producibility and cost in the U.S., Canada, and Europe, 1985.

3. Kandebo, Stanley W. "Researchers explore slush hydrogen as fuel for national aero-space plane" in Aviation Week & Space Technology, June 26, 1989, pp 37-38.

4 "NASA program investigates slush hydrogen", The Hydrogen Letter, August 1988, pp. 1, 4.

5. Boehman, Louis I., "Liquid hydrogen fuel simulant development", Proceedings of 29th Structures, Structural Dynamics and Materials (SCM) Conference 1988, pp 334-340.

6. "First plane flies exclusively on hydrogen", The Hydrogen Letter, June 1988, pp. 1, 4.

7. Nelson, Howard G., "Hydrogen and advanced aerospace materials", Sampe quarterly, October 1988, pp. 20-23.

8. Biggiero, G. and Borruto, A., "Mechanical behavior of sorbitic and austenitic steels in strongly hydrogenating environments: relevant use prospects", Int. J. Hydrogen Energy, Vol. 13, No. 12 1988, pp. 743-747.

Susan Giegerich translated this paper from German into English, prepared the final manuscript and looked after the figures. I thank her cordially.

POLLUTANT EMISSIONS OF EXISTING AND FUTURE ENGINES FOR COMMERCIAL AIRCRAFT

H. Grieb and B. Simon

MTU Muenchen GmbH

Abstract

This paper first describes the existing and the anticipated requirements to be met by commercial aircraft engines and discusses them with a view to the development work and measures considered necessary to effectively protect the earth's atmosphere.

Then the present level of pollutant emissions attained with today's engines is explained taking the CF6-50C engine as an example. Furthermore this paper outlines the development potential of future engines for subsonic commercial aircraft with respect to a reduction of the specific fuel consumption and the specific pollutant emission. In this connexion, possible future emission levels attainable with kerosine, methane and hydrogen are compared.

The reduction of specific fuel consumption and specific pollutant emission achievable with the aid of new technologies are set against the expected development in air traffic and the resulting tendency towards an increase in the fuel consumption by commercial aircraft worldwide. This comparison shows that a perceptible lessening of the impact on the atmosphere in and above the tropopause is possible only in the long run provided the specific fuel consumption as well as the specific pollutant emissions of engines can be drastically reduced.

Contents

1) REQUIREMENTS

Over the past 30 years, that is to say since the coming of the jet age, commercial aviation has developed enormously, thanks to the great advances in economy, productivity and flight safety. A decisive factor in this progress has been the reduction in direct operating costs, made possible not least by the reduction in fuel consumption per passenger-mile. This has been accompanied by a reduction in noise levels, resulting from the gradual change-over by operators to quieter aircraft following the introduction of more stringent FAR noise regulations in 1969. And now, since 1982, we have the ICAO recommendations, accepted worldwide, concerning the limitation of pollutant emissions, even if - corresponding to the level of knowledge at that time - these cover only the phases take-off and landing as well as ground movements.

Considering the continuing growth in air traffic as well as the increase in knowledge about the effects of engine exhaust and pollutant emissions at great altitudes, specifically the troposphere and above, not considered in the past, the aerospace industry as well as the public are becoming increasingly aware of the need for a careful review of future developments in commercial aviation in order to ensure that due attention is paid to protecting the earth's atmosphere.

In the aero-engine industry this growing awareness of the need for conservationist policies will mean that in comparison with the previous goals of greater economy, productivity and safety, this aspect of the impact on the environment will be the guiding principle behind the efforts directed at technological progress, where two courses will be pursued, namely

- the indirect reduction of pollutant emissions by reducing fuel consumption, and
- the direct reduction of toxic emissions by the use of new combustion chambers without changing the fuel consumption.

The priorities of the development of jet engines for civil aviation are listed in figure **1**. A look at this figure shows how in every case, the priorities to date have been concerned with the question of safety; whereas in future, the protection of the earth's atmosphere and the environment will assume particular importance. Certainly, no development in this direction will be accepted that might be interpreted as running counter to the high standard of safety already achieved. Accordingly, the trends in the development of engine designs and component technologies are discussed below from the point of view of their suitability for contributing to the protection of the earth's atmosphere and the environment.

2) STATE OF THE ART IN ENGINE DEVELOPMENT

Based on the above requirements and their future development, the paper first takes a look at the situation regarding operational reliability, specific fuel consumption, noise and pollutant emissions.

Concerning flight safety from the point of view of operational reliability, the inflight shutdown rate is all-important. Expressed in these terms, the improvement in flight safety, comparing engines of the first decade of jet-powered flight and modern engines, is as follows:

	In flight shutdowns per 1000 flying hours	
	Engines in the period following type certification or introduction into service	Fully-developed engines
- in first decade of jet-powered flight	0.5 - 1.0	0.1 - 0.3
- modern engines	0.08 - 0.12	0.01 - 0.02

In other words, compared with the beginning of jet-powered flight, the number of inflight engine shutdowns has been reduced by a factor of 10, representing an extraordinary increase in flight safety.

The development over the past thirty years and expected trends over the next twenty years of the energy consumption, i.e. specific fuel consumption, of aero-engines is represented in figure **2**. Following the introduction in the eighties of the 3rd generation of turbofans, i.e. those with high bypass ratio, a new generation featuring the propfan offering a reduction in specific fuel consumption to the order of 12 - 18% can be expected to be introduced towards the end of the nineties. It has still not been decided if both contestants, the unshrouded versus the shrouded version, will be successful for reasons that have still to be discussed.

The advances made in noise reduction can be seen in figure **3**, which compares the limits set in 1969 according to FAR stage 2 with those set in 1978 according to FAR stage 3. In view of the increase in noise which will accompany the growth in air traffic, associated with intended extension of take-off and landings in the early morning and late evening, a further tightening of the regulations is being discussed. As from 1990, any aircraft which fail to meet the stage 2 limits will no longer be allowed to fly. The figure also makes clear the length of time required for the fleets worldwide to be converted to the new standard. This, of course, is determined to a large extent by the financial resources of the airlines.

The restrictions regarding pollutant emissions in keeping with the internationally-accepted ICAO recommendations have resulted in the developments and progress represented in figure **4**.

In considering these figures, it is striking how in contrast to unburnt hydrocarbons (UHC) and carbon monoxide (CO), which have been greatly reduced, there has been no diminution in nitrous oxides (NO_x) emission. This can be explained by the fact that whilst improvements have been made to combustion chambers themselves, the development of engines with higher overall pressure ratios has resulted in physical prerequisites more unfavourable for reducing the emission of NO_x.

The present situation concerning aero-engines and pollution during the various phases of a complete flight cycle can be explained by taking the General Electric CF6-50C as an example. This engine exists in a number of versions, which differ by up to 20% with respect to pollutant emissions. Qualitatively, however, the behaviour of these engines is essentially the same.

Comparable pollutant-emission figures are exhibited by engine families such as the Pratt & Whitney JT9D and Rolls-Royce RB211. Certain differences in comparison with the latest models, such as the CF6-80, PW4000, PW2000, V2500 or CFM56, exist only at idle, where the newer engines are characterized by lower emissions of unburnt hydrocarbons.

The accompanying figures illustrate the emissions of NO_x, UHC, CO and particulates for the phases take-off at sea-level, initial stage of climb, final approach and taxi. These phases are defined according to the ICAO LTO cycle (see figure 5). The emission figures for the respective phases are based on ICAO Exhaust Emission Data Bank information.

The values for cruise are considerably more difficult to measure, all the more so since the emissions are influenced by a number of factors such as altitude, speed, duration of flight, type of aircraft and aircraft loading. Furthermore, it is not possible to measure the emission figures for cruise on normal test facilities; instead, special altitude test facilities associated with correspondingly high costs must be used.

Thus, the emission figures shown in the figures for cruise (10.7 km, M = 0.85) do not apply to the same CF6-50 version used for the ICAO cycle data, but are estimates based on various information in the literature.

Figure **6** shows the pollutant emissions referred to fuel consumption (emission index EI). This parameter is most appropriate for assessing the combustors of various engines or engine categories. Typically, NO_x emissions are particularly high on take-off, whereas this applies to UHC and CO during taxi. At cruise, the emission of NO_x is approximately half as great as during take-off, and the UHC and CO emissions are negligible.

If we take the absolute fuel consumption for each phase of flight as illustrated in figure **7** and multiply by the emission index according to figure **6**, we obtain the emission rate per hour as shown in figure **8**. In this case, for example, the NO_x emission rate per unit of time at cruise corresponds to roughly 1/6 of the rate per unit of time at take-off.

A simplified flight cycle for a short domestic flight, consisting of the ICAO cycle plus 30 minute cruise phase, as shown in figure **5**, results in the pollutant emission figures for each phase illustrated in figure **9**. Because of lack of data, the final part of climb and initial approach, where the altitude is greater, have been added to the cruise phase, because in the first instance somewhat greater and in the second somewhat lower values than at cruise are expected. According to this, for the given flight cycle, approximately 2/3 of the NO_x emission occurs during cruise, and 80% of the CO emission occurs during taxi. UHC emissions occur almost exclusively during taxi.

Finally, figure **10** shows the emission of particulates in the various phases of flight. There are no data available for cruise. The emissions are shown as the smoke number. It lies well below the limit of visibility, which is in the region of SN = 22 for the CF6-50 engine, in all phases of flight.

3) TRENDS IN FUTURE ENGINE DEVELOPMENT REGARDING POLLUTANT EMMISSIONS

NEW PROPULSORS

In response to the oil crisis in the seventies, in the United States NASA initiated a series of aero-engine technology programmes, including the unshrouded propfan in particular. This concept represents the transfer of the swept-wing effect, aimed at reducing the drag at high flying speeds of modern aircraft, to the propeller, enabling the propfan-driven aircraft to attain flying speeds comparable to those of the jet aircraft, that is to say in the region of Mach 0.8. Whilst initially the simple single-shaft propfan was pursued, since the beginning of the eighties it has come to be realized that for various reasons the counter-rotating version is more economical and offers clearly superior operating characteristics (see figure **11**, bottom left).

Sponsored by the Federal Ministry of Research and Technology, starting in 1983, in conjunction with MBB, MTU thus carried out a study to establish if the said economic advantages of the unshrouded propfan over the turbofan exist in fact. The findings showed that the reduction in direct operating costs, attainable with the more favourable counter-rotating propfan, of a short to medium-haul airliner of the Airbus A320 class is marginal because any advantages offered by the considerable reduction in specific fuel consumption are largely offset by the higher installation weight and cost of the engine. The study revealed in particular that the installation of the unshrouded propfan on the wing is impracticable in view of the measures that would be required to soundproof the cabin.

As a compromise between the turbofan and unshrouded counter-rotating propfan, these findings gave rise to the counter-rotating shrouded propfan (CRISP), which combines the advantages of both engines (figure **11**). More or less at the same time, P&W commenced studies into a shrouded single-rotation propfan known as the Advanced Ducted Propfan (ADP). Both concepts have been investigated and compared with regard to their feasibility and economy. The studies resulted in mechanical designs such as can be seen in figure **12**. Concerning the two concepts, it should be noted that

- the ADP generally speaking makes for a simpler and thus more cost-favourable engine, which can thus also be regarded as being less prone to risks; whereas
- the CRISP is characterized by the fact that for a given frontal area, the specific fuel consumption is 3 - 5% better than that of the ADP and that the highly regarded gearless design of the conventional turbofan is feasible.

A detailed description and analysis of the shrouded propfan concepts in comparison with propfans and conventional turbofans is provided in Ref. [1].

Seen overall, the attainable reduction in specific fuel consumption amounts to 10 - 11% with the ADP, and 14 - 15% with CRISP. Experience shows that in comparison with the turbofan, the CRISP can be expected to bring savings in direct operating costs especially in long-haul operations, whereas the ADP will be cheaper in short- and medium-haul operations.

From this it becomes apparent that the new propulsor concepts ADP and CRISP are capable of making a not insignificant contribution towards the indirect measures for reducing the impact on the earth's atmosphere mentioned above.

FUTURE CORE ENGINES

Besides the described method of reducing the specific fuel consumption of new engines, which is based essentially on improving the propulsive efficiency by reducing the specific thrust, another and in a sense final possibility of reducing the specific fuel consumption lies in the development of more efficient core engines, i.e. gas turbines, used for providing the propulsor driving power.

This calls in particular for a further increase in the turbine entry temperature and overall pressure ratio, but also for an improvement in the component efficiencies. The present and expected course of development of the turbine entry temperature and overall pressure ratio is shown in figure **13**. As made clear in figure **14**, the relationship between the two is thermodynamically justified, where their statistical development corresponds pretty well to the technical and economic optimum in that in forsaking the aerodynamically and technically difficult attainment of the pressure ratio most favourable as far as fuel consumption is concerned, somewhat lower pressure ratios are selected at the expense of somewhat higher specific fuel consumption.

Moreover, the fact should not be overlooked that - as already discussed in connexion with figure **4** - a further increase in the overall pressure ratio would vitiate the accompanying efforts to reduce NO_x emission.

On top of this, any further increase in the turbine entry temperature represents a rise in specific core engine performance, associated with a decrease in core engine main dimensions for a given propulsor performance or size. The history over the past two decades and the forecast into the next century of the development of the core-engine share of the total mass flow and total thrust is illustrated in figure **15**. The decreasing share in mass flow in association with the increasing compression leads to extremely small duct heights at the compressor exit and in the high-pressure turbine, with the result that the retention let alone the improvement of present efficiencies becomes no simple matter.

The considerable influence of the component efficiency, pressure losses and cooling-air requirement on the specific core engine data becomes clear from the following table, in which the figures are referred to a state of the art turbofan:

	Effect on core engine (Max. cruise thrust; M = 0.80; H = 11 km)	
	Specific power	SFC
1% improvement in efficiency of all turbomachinery components	+ 4.5%	- 2.6%
1% reduction in total pressure losses	+ 0.4%	- 0.4%
1% reduction in HP turbine cooling-air requirement	+ 2.0%	- 0.5%

In view of the above, it becomes logical to consider combining highly favourable fuel consumption plus moderate overall pressure ratios with appropriate conditions for reducing NO_x emissions. The classic technical solution to the problem of achieving very good specific fuel consumption as well as favourable conditions for reduced NO_x

emission lies in the recuperated gas turbine engine with or without intercooler. The relatively low overall pressure ratio of the cycle featuring recuperator and intercooler in comparison with the conventional cycle is illustrated in figure **14**. Studies to date have shown that despite the greater design investment involved, the concept featuring intercooler and exhaust heat exchanger in general offers the following advantages in comparison with the version with heat exchanger alone:

- The specific power of the core engine is some 25 - 30% greater, meaning that its mass flow and/or dimensions are reduced correspondingly
- The specific fuel consumption is about 7% better
- Low-temperature cooling-air is available at the compressor exit
- Moderate gas temperature on entry into the heat exchanger even with higher turbine entry temperature

The bottom half of figure **16** shows the design of the CRISP with advanced technology core engine of conventional type, i.e. with high compression and no heat exchanger, corresponding to the state of the art of around the year 2000. In contrast, the top half outlines the CRISP with the same thrust but featuring an intercooled, recuperative core engine, representing the state of the art which will presumably not be attainable before the year 2010 however. In this arrangement, the intercooler receives its air from the propulsor, where the gain in thrust from the recovery of heat outweighs the thrust sacrificed through pressure losses.

The cycle data of the core engines according to figure **16** can be summarized as follows:

Engine	CRISP without heat exchanger	CRISP with heat exchanger
Possible introduction date	2000	2010+
Turbine entry temperature at take-off, hot day	1600 - 1700 K	1700 - 1800 K
Pressure ratio at max. cruise	38:1 - 50:1	26:1 - 28:1
Bypass ratio	25:1 - 27:1	30:1 - 34:1

On the one hand, it should be noted that the development of efficient core engines without heat exchanger is also a possibility with future turbofans. On the other hand it goes without saying that the realization of recuperative core engines depends on the availability of compact, light and highly-efficient heat exchangers. Thanks to the investigations it has been conducting since 1979, MTU already has acquired an excellent technological basis in this area, compare Ref. [2].

The specific fuel consumption expected of future CRISPs with conventional or recuperative core engines related to the important operating range is illustrated in figure **17**. From this it can be seen that the advantageous consumption of the recuperative core engine stems from the known favourable part-load characteristics of the recuperative gas turbine. This is made clear in figure **18**, which provides a comparison of the specific fuel consumption at part-load of the state of the art turbofan and the concepts of the future. This shows that in comparison with the turbofan of the same standard, the shrouded propfan with conventional core engine of the state of the art of the year 2000 is characterized by a reduction in fuel consumption of around 11% in the most important thrust range, and that for the year 2010 this figure is about 15%. The shrouded propfan with recuperative core engine of the year 2010+ is expected to give a further saving in the region of 13% over the propfan with conventional core of the same standard.

Based on a long-haul aircraft of the class of the Airbus A340, a weight situation as follows can be expected:

Take-off weight	260	t
Number of passengers	300	
Range	7000	NM
Cruising speed	0.82	Mach
Cruising altitude	31,000 -	39,000 ft
Thrust loading at take-off	0.25	daN/kg

Engine concept	CRISP Conventional core	CRISP Recuperative core
Mission fuel-load	109 t	95 t (-13%)
Engine weight (4 engines)	4 x 3.0 = 12 t	4 x 4.0 = 16 t (+33%)
Total propulsion weight (engines + fuel)	121 t	111 t
Weight saving	-	10 t
Consequences for aircraft:		o More freight
	and/or	o Greater range
	and/or	o Reduced weight
	and/or	o Improved wing design

As a result, a clearly improved aircraft with reduced take-off weight can be expected. However, as attractive as the reduction in fuel consumption over the long term may be, it must not be overlooked that the development of a recuperative core engine will be very difficult in view of the aspects weight and reliability. Moreover, gaining the acceptance of the operators will not be so simple.

POLLUTANT EMISSIONS

It is clear that the shrouded propfan described above will play its part in the indirect approach to protecting the atmosphere, i.e. through reduced fuel consumption. With the core engines also described above on the other hand, the problem is approached both directly and indirectly in that in addition to the improved fuel-consumption characteristics, the redesign of the combustion chambers in particular can be expected to make a major contribution. Initially it is assumed that the measures envisaged so far, namely as shown in figure **19**, will apply equally to both core-engine concepts. However, since the third step is valid only up to certain operating pressures, it will probably be most effective in conjunction with the recuperative core engine.

In any case as made clear in above discussion on future core engines, the development of conventional core engines aimed at greater economy will result in higher pressure ratios, meaning that, as shown on the left of figure **20**, the physical conditions necessary for reducing NO_x emissions through the above measures will become significantly more difficult. Consequently it is possible that under certain circumstances the further development of conventional core engines with higher overall pressure ratios will be hindered or even prevented because of their inability to comply with possible future regulations limiting the NO_x emissions. Figure **20** also makes it clear that if the trend towards higher pressure ratios continues, the situation can be improved only through the second and third measures, shown in figure **19**. But it is still uncertain if they can be realized in practice and if they will be approved. In contrast, with the recuperative core engine shown on the right in figure **20**, the lower pressures and temperatures on entry into the combustion chamber mean that significantly reduced NO_x emissions are to be expected and that these emissions will remain below the present levels even if the engines are further developed to achieve greater economy.

The methods of reducing pollutant emissions and the problems expected in this connexion are described next.

4. COMBUSTOR CONCEPTS FOR LOW POLLUTANT EMISSIONS

CONVENTIONAL COMBUSTION CHAMBERS

The design of combustion chambers for gas-turbine engines is determined by two factors. Firstly the engine load is controlled by varying the turbine entry temperature and thus the overall fuel-air mixture ratio in the combustion chamber. Secondly, this overall mixture ratio is so lean in all stages of operation that stable combustion would not be possible. To enable an engine to be controlled accompanied by simultaneous stable combustion, the combustion chamber consists of zones with different temperature levels (figure **21**).

The first or primary combustion zone is supplied with a portion of the air from the compressor and at full load is operated virtually stoichiometrically, whilst at idle the equivalence ratio i.e. the fuel-air ratio related to the stoichiometric ratio, which is an indication of the richness of a mixture, drops to about 0.5. By this means safe engine control is achieved.

If on the other hand all of the compressor air were to be mixed with the fuel, the combustion chamber flame would be extinguished at part-load operation (figure **22**) and control would be impossible.

However, as a consequence of this method of engine control large amounts of NO_x are produced under full-load operating conditions, since NO_x formation is greatest in the region of stoichiometric combustion. The formation of CO and UHC, which sets in at low temperatures close to the extinction limit, can be largely avoided, as can the smoke formation. Thus, typically, the NO_x and smoke emission of a jet engine is greatest on take-off, whereas the emission of CO and UHC is greatest at idle and low engine load conditions.

Whilst CO and UHC emissions have been drastically reduced in modern jet engines, the emission of NO_x has even increased. The reason is to be found in that the overall pressure ratio of jet engines is continually increasing, resulting in a virtual doubling of the NO_x emission index over the past 20 years.

LEAN COMBUSTION

In order to reduce NO_x emission, the equivalence ratio in the primary zone, which is responsible for the emission, must be modified.

Taking another look at figure 22, which shows the combustion temperature plotted as a function of the primary zone equivalence ratio, putting it simply, one can distinguish three zones in which pollutant formation occurs. At low temperatures and marked lean conditions, CO and UHC are formed. In the region around stoichiometric combustion, marked NO_x formation occurs, and smoke starts to form with rich mixtures above an equivalence ratio of 1.6.

There are two narrow windows between these regions in which pollutant formation is low. With conventional combustion chambers the full-load operating point of the primary combustion zone lies in the region of maximum NO_x formation. This is deliberate to provide a wide range of engine control.

In order to reduce the NO_x formation, the operating point can be shifted into one of the windows where the formation of pollutants is the smallest in the lean or rich region. Shifting of the operating point into the lean region results in corresponding lean combustion (figure 23). In practice this is achieved by feeding the whole of the compressor air to the primary combustion zone, and not just a portion of it. In this case, it is very important that the combustion mixture is homogeneous, otherwise local hot-spots occur, which results in the formation of NO_x.

There are two ways in which the required homogeneity can be achieved. Firstly, the fuel can be premixed and prevaporized externally before the mixture enters the combustion chamber. Secondly, this premixing section can be omitted and the fuel can be injected directly into the primary combution zone and premixed there. However, there is a risk that the fuel will ignite before the premixing process has been completed.

It must be borne in mind that the range of control of these combustion chambers is very small in comparison with conventional types and is no longer adequate for purposes of engine control. Hence, measures must be taken to widen the range of control, whilst maintaining the desired optimum equivalence ratio in the primary combustion zone, as described below in greater detail.

LEAN COMBUSTION WITH PREMIXING AND PREVAPORIZATION

The NO_x emitted by modern engines is thermal NO_x, meaning that lowering of the temperature by lean combustion in the primary zone is a very effective method of reducing the emission of NO_x. For this, the fuel must be well mixed with the air. In the most favourable case, the fuel should be completely vaporized and the fuel and air ideally mixed before entering the primary zone of the combustion chamber. This requires the introduction of an external premixing section in which the air is premixed and prevaporized.

Premixing sections are used with success in stationary gas-turbine installations. But it must be remembered that here the combustion chamber pressures and temperatures are relatively low and that the residence time in the premixing section, allowable in relation to spontaneous ignition, is very long (< 1 s). By contrast, under the conditions prevailing in aero-engines, the acceptable residence time is no longer than about 6 ms, compare figure 24.

Figure **25** shows the degree of vaporization expected under these circumstances. With a combustor entry temperature of 750 K, the degree of vaporization is in the region of 85%. However, bearing in mind that the allowable residence time in the premixing section according to Ref. [3] is subject to a relatively large degree of uncertainty, it only applies to the premixing section described there.

Thus, in aero-engines a safety factor of about 3 is called for. As a result of which, the residence time is cut to about 2 ms. Under these conditions the degree of vaporization reduces to 55%. It must also be borne in mind that in modern aero-engines at take-off, the pressures and temperatures at combustor entry are very high (30 bar, 850 K), meaning that the use of lean combustion with premixing and prevaporization is restricted to cruise with today's engines and fuels. This has been verified experimentally, where companies such as General Electric have great experience. It has tested combustion chambers operating according to the principle of lean combustion with external premixing under various conditions, where damage to the premixing section occured at a pressure of approximately 20 bar and a combustor inlet temperature of approximately 830 K. It was not possible to clarify if this damage was the result of spontaneous ignition or flashback, see Ref. [4]. Regarding to reliability and safety, lean combustion with external premixing and prevaporization is limited in use.

To avoid the dangers inherent in external premixing, lean combustion with internal mixing is also possible. But there is then the danger of the fuel ignition before a sufficient degree of mixing has occured, resulting in high NO_x emission. In addition, ignition of the fuel droplets can occur before they have been completely vaporized and have entered the combustion chamber. This also leads to increased NO_x emission.

Finally, marked inhomogeneities in the primary zone region also lead to increased CO emission. For this reason, the potential for reducing the emission of NO_x with lean combustion and internal mixing is relatively small. Reduction of around 30 - 50% was attained according to Ref. [5]. In contrast, the CO and UHC emissions were very low.

COOLING REQUIREMENTS

There is yet another aspect which limits the use of the lean combustion principle over the mean term. This is the fact that with increasing turbine entry temperature, the air required for combustion increases the more, the greater is the degree of lean combustion desired. As illustrated in figure **26**, the amount of cooling air available decreases proportionately. Consequently, with a primary zone temperature of 2000 K and a combustion chamber exit temperature of 1800 K, the proportion of air available for cooling is only about 20%. Whilst it is true that the cooling air requirement with lean combustion is not as great as with stoichiometric combustion, this reduction is not nearly as great as that which occurs in the amount of cooling air available. Hence, it is necessary either to find methods of cooling which require less air or allow a higher combustion temperature, which again leads to increased NO_x emission, or higher combustor liner temperatures.

RICH-LEAN COMBUSTION

If the operating point in the primary zone is shifted into the window in which low emission occurs in the rich region, a second, lean recombustion stage must be added after the primary, rich combustion zone. This is correspondingly known as rich-lean combustion (figure **27**).

Transition from the first to the second zone is effected by the rapid addition of air. Once again, the range of control is much smaller than with conventional combustion chambers and is far from adequate for engine operation.

The emission of NO_x is at its minimum with an equivalence ratio of around 1.6. However, under these conditions increased CO, UHC and smoke emission occurs. The emission of smoke increases with increasing equivalence ratio and increasing combustion chamber pressure. The threshold at which the limits of smoke emission are exeeded is frequently below the optimum equivalence ratio for NO_x emission. This applies the more, the higher is the combustion chamber pressure. The relationship between smoke and NO_x emission at high pressures and temperatures corresponding to take-off conditions has not been sufficiently studied to date.

The mixture problem is of importance for rich-lean combustion for two reasons:

- In a similar way as with all combustion chambers, for the rich stage it is important that the mixture of the fuel and air be as homogeneous as possible in order to reduce the tendency towards smoke formation and for NO_x emission.

- More important however is the rapid admixing of air during the transition from the rich to the lean zone, because stoichiometric mixture processes are involved, resulting in the formation of thermal NO_x. It is then vital that as homogeneous as possible a mixture in the lean zone is reached as speedily as possible.

How successful will be the reduction in NO_x emission accompanied by satisfactory CO, UHC and smoke emissions by means of rich-lean combustion after all the above problems have been solved, is very uncertain. From the theoretical point of view, rich-lean combustion could be as effective as lean combustion with premixing. This is illustrated in figure 28, which shows analytical results on the influence of the equivalence ratio and residence time on the total fixed nitrogen, i.e. mainly of NO_x. Considering that since both the smoke problem and the need for rapid mixing represent additional difficulties, a reduction in NO_x emission in the order of 60 - 70% is fairly realistic.

Similar to the lean combustion chamber, cooling is also a problem with the rich-lean combustion chamber, although in this case the problem is confined to the special conditions obtaining in the rich zone. The use of film-cooling is not permitted here, since it would result in excessive oxygen and thus the formation of thermal NO_x. Only convection cooling from the outside is permitted. The question of whether this is adequate given the high pressure conditions, or if new heat-resistant materials such as ceramics will have to be used still needs to be answered.

CONTROL OF LOW-EMISSION COMBUSTION CHAMBER

In principle there are two methods of decoupling the fuel-air ratio at the end of the combustion chamber from the quivalence ratio in the primary combustion zone, namely by fuel staging or air staging (figure 29). Fuel staging is suitable only for lean combustion, whereas air staging (variable geometry) is required for rich-lean combustion.

With fuel staging, different combustion zones are provided for idle and full-load operation, where these zones can be arranged axially or radially (figure 29). At idle, burning takes place only in the pilot zone, with the main zone being activated as the engine load increases. Lean combustion under various operating conditions can be achieved by matching the airflow for the pilot and main zones accompanied by variation of the distribution of the fuel to the two zones. However, it needs to be borne in mind that unsteady control will occur, especially when switching in the main zone, and there will not always be an optimum mixture ratio in both zones.

More effective from the point of view of combustion chamber control is air staging, where the optimum mixture ratio in the primary combustion zone is achieved by varying the proportion of air in the mixture. This calls for a control mechanism capable of varying cross-section responsible for the supply of air to the primary zone, where this variation must be compensated by a corresponding change in the mixing zone. By this means it is possible to maintain the desired mixture ratio in the primary zone. However, it has to be said that the presence of moving parts in the vicinity of the combustion chamber with its high pressures and temperatures involves considerable risk. Hence, particular attention must be paid to reliability and airworthiness at the design stage.

5) COMBUSTOR CONCEPTS FOR ALTERNATIVE FUELS

GENERAL

Regarding the use of alternative fuels as a means of reducing the ecological impact of future aircraft, attention is currently concentrated on hydrogen and, to a lesser extent, on methane.

Irrespective of emission, there is no question that hydrogen is preferable to all other fuels when it comes to hypersonic flight or space flight, not just because it is capable of providing the greater specific thrust required in these applications, but also because it offers the much sought-after greater cooling capacity. However, with subsonic aircraft, which are likely to predominate for some time to come, these properties are not required.

Moreover, as early as the fifties, hydrogen was envisaged as a fuel for high-altitude aircraft thanks to its superior combustion stability at low pressures. Bearing the high life requirements of modern engines and today's stringent certification regulations in mind, the solution to the problem of hydrogen embrittlement of highly-stressed hot-section components, encountered in conventional jet engines fuelled by hydrogen, will not be simple, but should not be impossible.

Since at least the seventies it has been known that for a given passenger-carrying capacity and range, subsonic airliners propelled by hydrogen will be significantly lighter, but more spacious. This calls for totally new designs. The underlying reason lies in the significantly lesser weight but greater volume referred to equivalent combustion energy of liquid hydrogen in comparison with kerosine, as shown in figure 30.

Since these earlier studies it has been quite clear that the significantly higher specific cost of liquid hydrogen left no hope of greater or even comparable economy of hydrogen-propelled airliners versus kerosine-propelled types. Moreover, the costs of setting up the infrastructure necessary for the introduction of liquid hydrogen as well as the solution to problems associated with aircraft safety were further arguments against this new breed of aircraft, but these arguments are well-enough known and need not be gone into here.

Furthermore, the emission characteristics of hydrogen in comparison with kerosine were already a topic for discussion as early as the seventies, when the problem of pollution in and above the tropopause was at least seen in its true perspective. However, the main considerations behind the choice of hydrogen as an alternative to kerosine were questions of economy and safety.

Although it is now appreciated that NO_x emissions can probably be kept very low with methane, the use of methane-propelled aircraft instead of kerosine-propelled ones was not a topic for discussion in those days. This was because with regard to economy this fuel did not offer any visible advantages that would have justified corresponding development. In fact, if anything, there would have been disadvantages.

CONVERSION OF COMBUSTOR CONCEPTS

Of late, hydrogen has again come into consideration as a fuel for commercial aircraft because of its supposed lesser impact on the earth's atmosphere at great altitudes, namely in and above the tropopause. This section, therefore, will be concentrated on the combustion of hydrogen and the discussion of the problems expected from this in comparison with the combustion of kerosine and methane. This situation is illustrated in figure **31**, which gives a qualitative comparison of the properties of the fuels under discussion, which are of relevance with regard to combustion or NO_x emission so critical at great altitude.

In principle, the most effective method of reducing the emission of NO_x is to be found in lean combustion in association with premixing. However, because of the danger of spontaneous ignition, with kerosine this process can be used only at low pressures and temperatures (at cruise, for example). When hydrogen or methane is used, however, this risk of spontaneous ignition with premixing is reduced. Conversely, the risk of spark ignition as a result of flashback is low with kerosine and methane. Because of hydrogen's extremely high flame-propagation velocity, however, the danger of flashback from the primary zone of the combustion chamber into the premixing zone is much greater with hydrogen than with kerosine and methane. Hence, the premixing concept cannot be used in hydrogen-fuelled engines.

It must further be borne in mind that, in contrast to Kerosine and methane, hydrogen burns immediately after entering the primary zone wherever it comes into contact with air, although extremely different equivalence ratios can occur locally at the entrance. The extremely high convective mixing rate and molecular diffusion capability of hydrogen in air will certainly dampen this effect, but cannot eliminate it altogether. Hence, one will have to reckon with the following, at least so long as today's injection systems are still in use:

- That local variations in the equivalence ratio where burning occurs in the primary zone will be greater with hydrogen than with kerosine, with the result that the desired conditions of lean combustion will indeed prevail in general, but not universally; and

- that because of its very wide ignition limits, hydrogen will also burn in areas where it is not supposed to, for example close to walls, resulting in thermal problems.

This situation is adressed by figure 32, which gives a survey of the ignition limits of the fuels at different pressures, where the extremely low lean ignition limit of hydrogen is striking.

In view of the fact that equivalence ratios of around unity probably cannot be avoided locally in the primary zone with hydrogen, and that the maximum possible combustion temperature of hydrogen is greater than that of kerosine and methane*, an unfavourable effect on NO_x emission is to be expected with hydrogen, at least for as long as today's injection systems are in use.

In contrast, it is to hydrogen's advantage that because of its very special properties, the combustion chamber can be much smaller in cross section and be shorter than combustion chambers for kerosine. This in turn means that the residence time of the combustion gases will be reduced and, since the rate of NO_x-formation under the conditions in question is more or less proportional to the residence time, the NO_x emisson will also be reduced correspondingly.

EXPERIMENTAL RESULTS

Experiments carried out to date with a combustion chamber burning kerosine, propane** and hydrogen as illustrated in figures 33 and 34 have demonstrated that in the present case, that is to say with the same combustion chamber configuration and the same aerodynamic and thermodynamic conditions, hydrogen produces by far the greatest NO_x emission, compare Ref. [6]. But it must be pointed out that

- a conceivable injection system developed specially for hydrogen and designed for particularly intensive mixing of air and hydrogen after the entry to the primary zone, and
- the correct dimensioning of combustion chambers for hydrogen, in other words for short residence times,

can contribute considerably towards reducing the NO_x-formation rate.

From the foregoing it should therefore not be concluded that the emission of NO_x is necessarily greater with hydrogen than with kerosine or propane (methane).

Apart from the combustion-technology aspects discussed above, it should be mentioned that concerning NO_x-formation, hydrogen's higher maximum combustion temperature in air associated with temperature restriction in the primary zone necessitates a lower equivalence ratio than with kerosine. Consequently, there will be less cooling-air available in the combustion chamber with hydrogen in comparison with kerosine, setting greater demands with regard to materials technology. However, this can be offset in that the ther-

* At entry conditions of p = 1 bar and T = 850 K, the max. attainable temperature is 2600 K with hydrogen as opposed to 2480 K with kerosine and 2530 K with methan.

** Propane (C_3H_8) is considered to behave similarly as methane (CH_4)

mally-stressed area of combustion chambers for hydrogen can be kept smaller than in those for kerosine.

An experimental programme concerning hydrogen- and methane-propelled aircraft, recently carried out in the USSR, is reported on in Ref. [7], where the most interesting aspects would appear to be the practical experience gained with cryogenic fuels.

OUTLOOK CONCERNING POLLUTANT EMISSIONS

Figure 35 shows the expected emission figures for hydrogen, kerosine and methane for combustion of the same energy-content under the same aerodynamic and thermodynamic conditions. An important consideration here is that the emission of water, which is considered to be particularly critical at great altitude, is noticeably greater with hydrogen than with kerosine or even methane. On the other hand, the lower emission of CO_2 in the case of methane and even the absolute avoidance of CO_2 emissions by hydrogen-fuelled engines is of no relevance, since even the absolute avoidance of CO_2 emissions at any altitude would be virtually meaningless in view of minimal role played by air traffic in the total emissions of CO_2 attributable to man.

As far as the emission of NO_x is concerned, figure 35 indicates that greater progress is to be expected with hydrogen than kerosine, since - as explained earlier - in addition to lean combustion, advantage can be taken of the shorter residence time when hydrogen is used as the fuel.

Apart from the fact that the starting situation is already favourable concerning methane, the prospects for this fuel are particularly propitious, since it would seem that lean combustion can be combined with premixing and prevaporization, thanks to methane's superior properties, illustrated in figure 31.

Finally, a comprehensive qualitative assessment of the aviation fuels under review, such as that represented in figure 36 indicates that in the light of today's knowledge, hydrogen's purported particular environmental friendliness is questionable.

6) LONG TERM TREND IN POLLUTANT EMISSIONS

It is certain that air traffic will continue to grow and more knowledge will be gained about the effects of engine exhaust and pollutant emissions at great altitudes, specifically the tropopause and above, which is a matter we have tended to neglect in the past. Consequently, the aerospace industry as well as the public are becoming increasingly aware of the need for a careful review of future developments in commercial aviation in order to ensure that due attention is paid to protecting the earth's atmosphere.

The impact on the atmosphere is illustrated as listed below, which shows the percentage of fuel burnt in the critical altitude range, based on Lufthansa figures for the year 1988. It can be assumed that analysis of air traffic worldwide would show similar results.

o Tropopause altitude (seasonal average) ranges between 8 km (at poles) and 15 km (at equator)

o Tropopause altitude in summertime is generally higher than in wintertime

- With the Lufthansa fleet 17 - 20% of total fuel is burnt in or above the tropopause
- Medium and long range flights below the tropopause would result in fuel burn increase of 6 - 8%

Figure **37** shows the expected development in air traffic worldwide. It can be seen how the volume of traffic is likely to double between 1988 and 2008, that is to say within 20 years.

In contrast to this enormous growth in air traffic, the potential fuel-savings achievable with subsonic aircraft are shown in figure **38**. According to this figure a reduction in fuel consumption per passenger-mile to the order of 45 - 50% is to be expected, thanks to improvements with regard to aircraft weight and drag as well as specific fuel consumption of the engines.

As mentioned in Section 1 technological developments in the aviation industry will no longer be able to be concentrated solely on the aspects safety and economy, but will have to pay particular attention to the effect they have on the earth's atmosphere. It thus goes without saying that the optimum answer lies in developments which will have a reduced impact on the atmosphere in addition to bringing greater economy.

To clarify the long-term situation regarding the emission of CO_2, H_2O and NO_x as a result of air traffic, the development possibilities with aircraft and aero-engines, discussed above, are summarized in figure **39**, which also shows the availability expected with the successive introduction into service of the aircraft and engines.

Further, the modular assumption is made that a new generation of aircraft or engines with appreciably smaller thrust requirement per passenger and lower thrust-related specific fuel consumption, as well as significantly-reduced pollutant emissions, will be available for introduction every 10 years. It is borne in mind that, as illustrated in figure **40**, experience shows that the introduction of a new generation of aircraft or engines (e.g. generation B) will proceed slowly at first, but will increase later, whilst in the first part of the same period predominantly aircraft of the previous generation (e.g. generation A) will be supplied to customers. Also based on current experience, it is further assumed that, as shown in figure **41**, aircraft of a certain type will be increasingly phased out after 16 years service, meaning that a maximum service time of 32 years can be expected.

A summary of these modular assumptions, which of necessity represents a considerable simplification in comparison with the highly-complex reality, produces a trend in the emission of pollutants as illustrated in figure **42**. From this it becomes immediately apparent that despite the assumed technological progress and its gradual introduction as per figures **39** and **40**, the emission of CO_2 and H_2O will continue to increase for decades as a consequence of the continued growth in air traffic. In contrast, it seems likely that the emission of NO_x, profiting from both improved fuel-consumption and from improved combustion chambers, will reach its maximum sooner, followed by a more rapid decrease.

The emission trend according to figure **42**, which of course is of a hypothetical nature, shows that even with the intensive development of new technologies and their gradual introduction, the impact on the atmosphere caused by emissions will continue to increase over the medium term, assuming that air traffic grows as prophesied.

Looking at figures 35 and 42, it should be emphasized that in comparison with kerosine, the use of hydrogen as an aviation fuel will not bring about any improvements with regard to NO_x emission, and that the greater emission of H_2O will even worsen the greenhouse effect. Moreover, it should be stressed that the question of whether the increase in the CO_2-content of the atmosphere associated with the burning of oil and the other fossil fuels will be kept within acceptable limits over the long term or not will depend not on aviation but on the earthbound consumer.

7. OUTLOOK FOR FUTURE AVIATION FUEL SUPPLIES

As far as specific energy costs are concerned, hydrogen will remain more expensive than kerosine as long as there are unlimited supplies of oil from conventional sources.

Here it should be pointed out that the oil supply situation can be expected to change drastically in the course of this decade, as illustrated in figure 43. In particular, the most important sources, namely those in the USA and USSR, will run out before the end of the century. This means that growing importance will be attached to the supplies in the Middle East with the associated political risks and possibility of price increases, the more the capacities of the Middle East sources can be increased significantly. Certainly the West has practically inexhaustible supplies in the form of oil shale, etc. However, exploitation will be very costly, resulting in higher oil prices as shown in figure 44. Over the long term, therefore, it will be worthwhile to keep an eye on how the relationship between the price of hydrogen and kerosine develops.

Bearing this background in mind, it seems reasonable to state that reducing the amount of fuel consumed by commercial aircraft will grow in importance, not just with regard to environmental considerations, but also because of the economical aspects.

8 CONCLUDING REMARKS

In view of the described possibilities of further reducing the emission of NO_x, coupled with the expected growth in air traffic, it becomes clear that even with the most intensive technological progress and the introduction of a new generation of aircraft and engines within the restraints of the funds available for development and procurement, the pollution of the atmosphere by aircraft can be expected to continue to increase for some time to come. Moreover, present knowledge indicates that the question of using hydrogen instead of kerosine as an aviation fuel is of secondary significance, even if a distinction has to be made concerning the arguments about the emission of NO_x, H_2O and CO_2.

The unhindered introduction of new technologies aimed at reducing the impact on the atmosphere can be expected only when they also result in economic advantages. A possible example would be the indirect alleviation of the impact on the atmosphere by reducing fuel consumption. Furthermore, the new technologies currently being developed for directly reducing pollutant emissions, particularly NO_x, will decisively contribute to satisfying possible future pollutant emission regulations above and beyond the existing ICAO recommendations.

In every case, however, the effects of the emissions on the atmosphere, especially at and above the tropopause, will have to be clarified. Consequently, the researchers everywhere must coordinate their efforts to meet the challenge of determining the effects of aircraft emissions on the earth's atmosphere and the global consequences.

ACKNOWLEDGEMENT

The authors wish to thank Herr Dipl.-Ing. Hans Gronewald for providing information about the introduction of aircraft and engines into service and about commercial operations.

REFERENCES

Ref. 1 Grieb H., Eckardt D.:
Propfan and Turbofan - Antagonism or Synthesis
ICAS Congress, 1986, London

Ref. 2 Pellischek G.:
Compact High-Performance Heat Exchangers for ICR Aircraft Gas Turbines
European Propulsion Forum, 1990, Cologne

Ref. 3 Spadaccini L.J., TeVelde J.A.:
Autoignition Characteristics of Aircraft-Type Fuels
Combustion and Flame 46: 283-300 (1982)

Ref. 4 Ekstedt, E.E.:
Advanced Low Emissions Combustor Program
AIAA-87-2035

Ref. 5 Sokolowski D.E., Rohde J.E.:
The E^3 Combustors: Status and Challenges
NASA TM 82684 (1981)

Ref. 6 Anderson, D.N.:
Emissions of Oxides of Nitrogen from an Experimental Premixed Hydrogen Burner
NASA TM X-3393, 1976

Ref. 7 Sosounov, V.
Some Aspects of Hydrogen and Other Alternative Fuels for Application in Air-Breathing Engines
ISOABE, IX International Symposium on Air-Breathing Engines, 1989, Athens

Priorities concerning Engine Requirements for Commercial Aircraft

Criterion	Requirement	Current: Long-haul	Current: Short/Medium-haul	Future (after 2000): Long-haul	Future (after 2000): Short/Medium-haul
Safety	- Reliability - Operation in all climatic zones - Protection against FOD			As before	
Profitability (low operating costs)	- Fuel consumption - Engine price - Maintenance costs			As before	
Protection of atmosphere/environment	- Fuel consumption - Pollutants - Noise	–	–		
Service flexibility	- Ease of installation - Thrust-enhancement potential - Weight and dimensions			As before	

Priority: Particularly important Very important Important

Fig. 1

Development History of Specific Fuel Consumption of Engines for Commercial Aircraft Cruise M=0.80; H=11 km

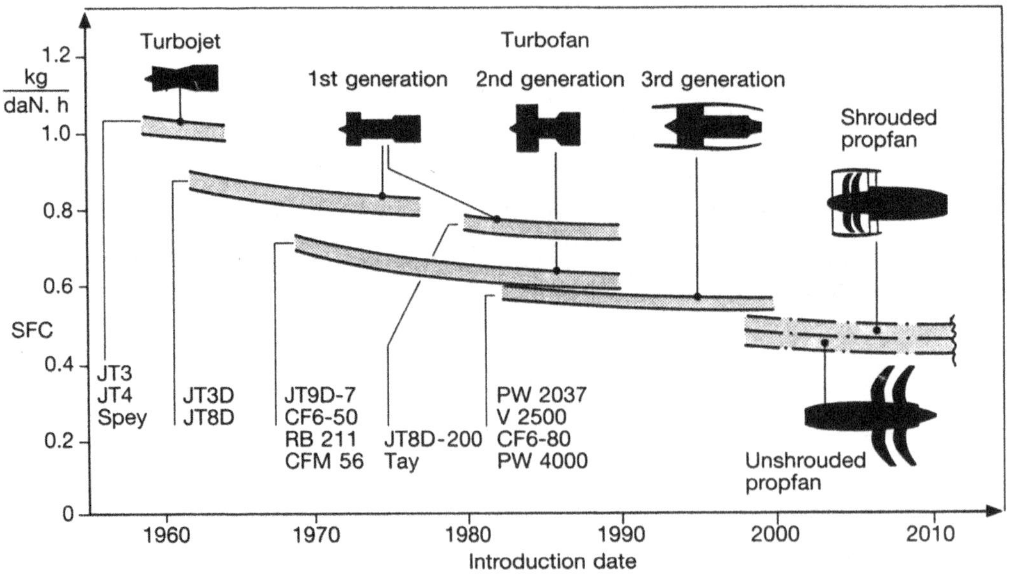

Fig. 2

Advances in Noise Reduction

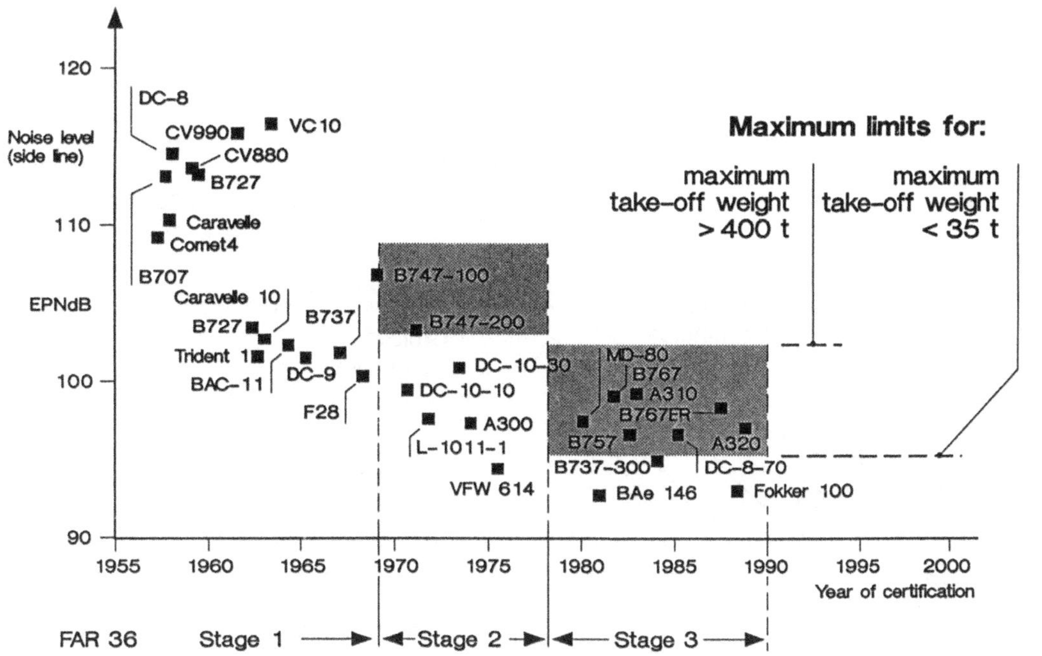

Fig. 3

Development History of Emissions of Various Aircraft Engines according to ICAO cycle

Source: ICAO Exhaust Emission Data (Bank)

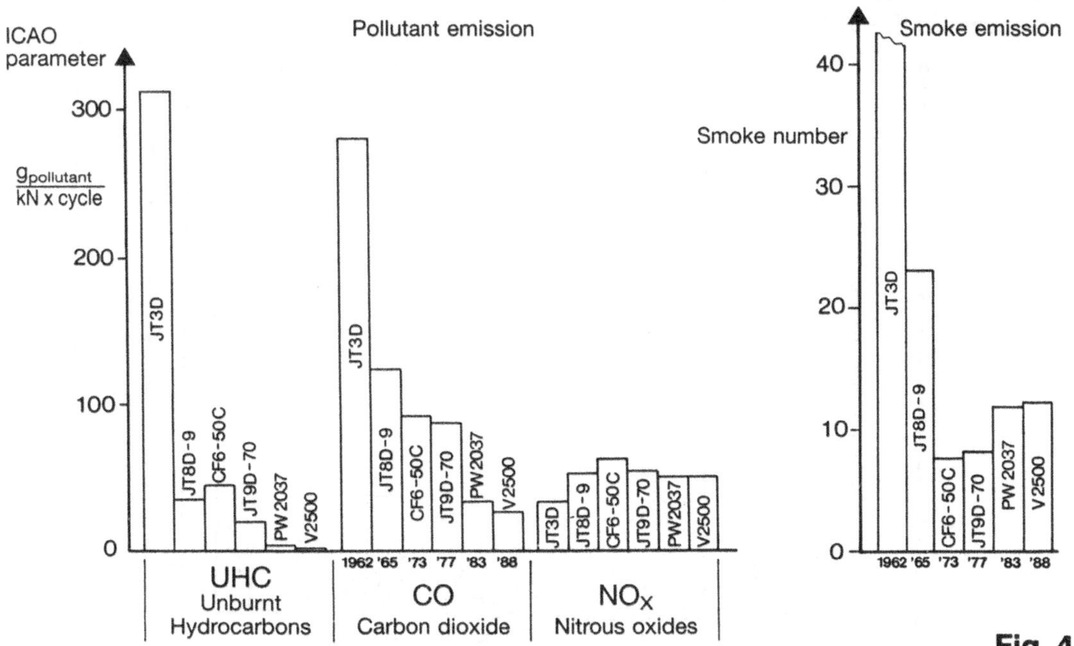

Fig. 4

Simplified Flight Cycle
(ICAO, LTO Cycle + 30 Min. Cruise)

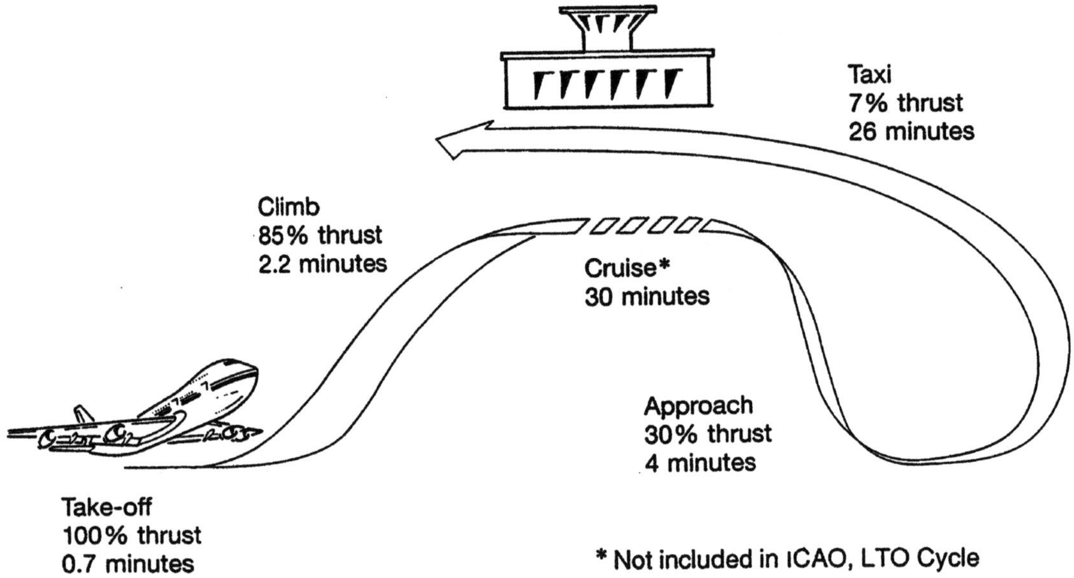

Fig. 5

Amount of Pollutants Referred to Fuel Consumption (Emission Index) for Various Phases of Flight Engine CF6-50C

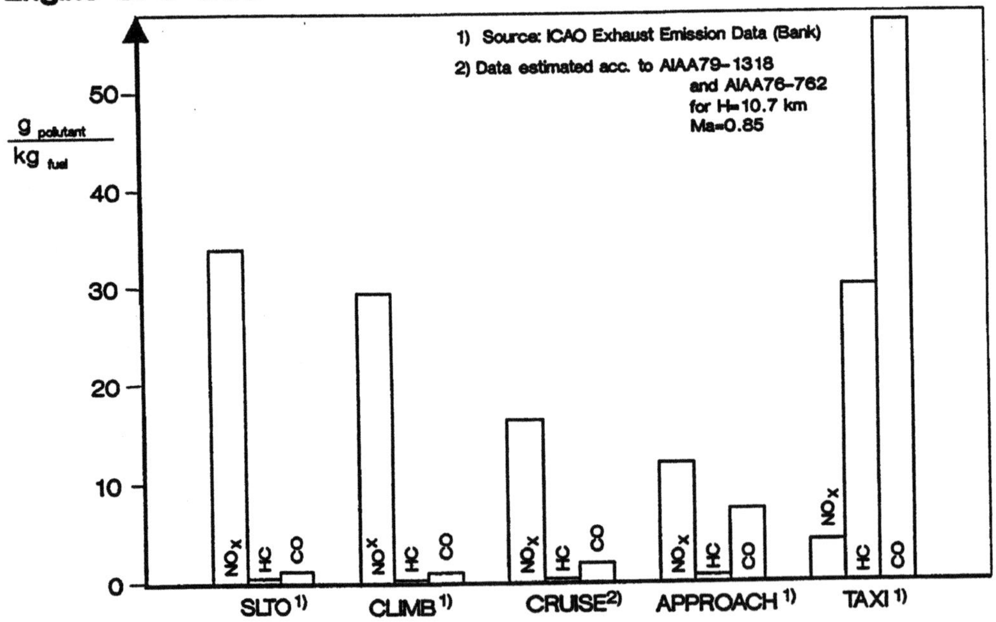

Fig. 6

Fuel Consumption for Various Phases of Flight
Engine CF6–50C

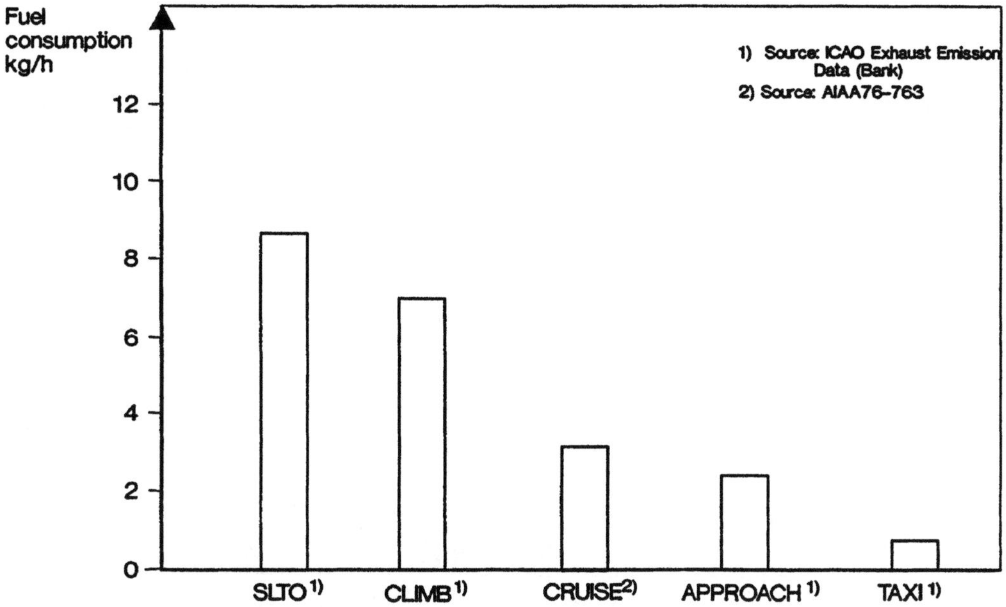

Fig. 7

Pollutant–Emission Figures for Various Phases of Flight
Engine CF6–50C

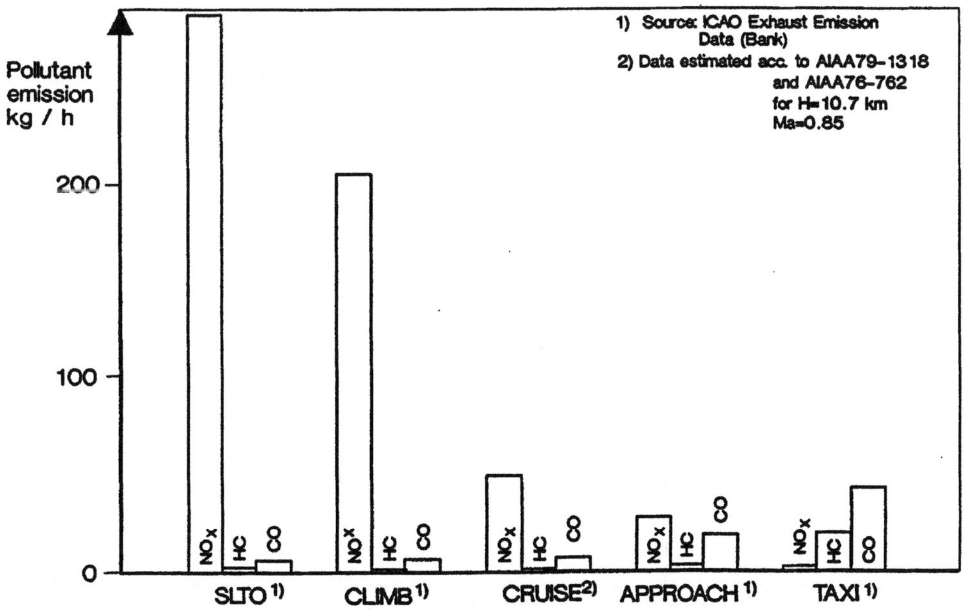

Fig. 8

Pollutant Emission for Each Phase of Flight
Engine CF6–50C

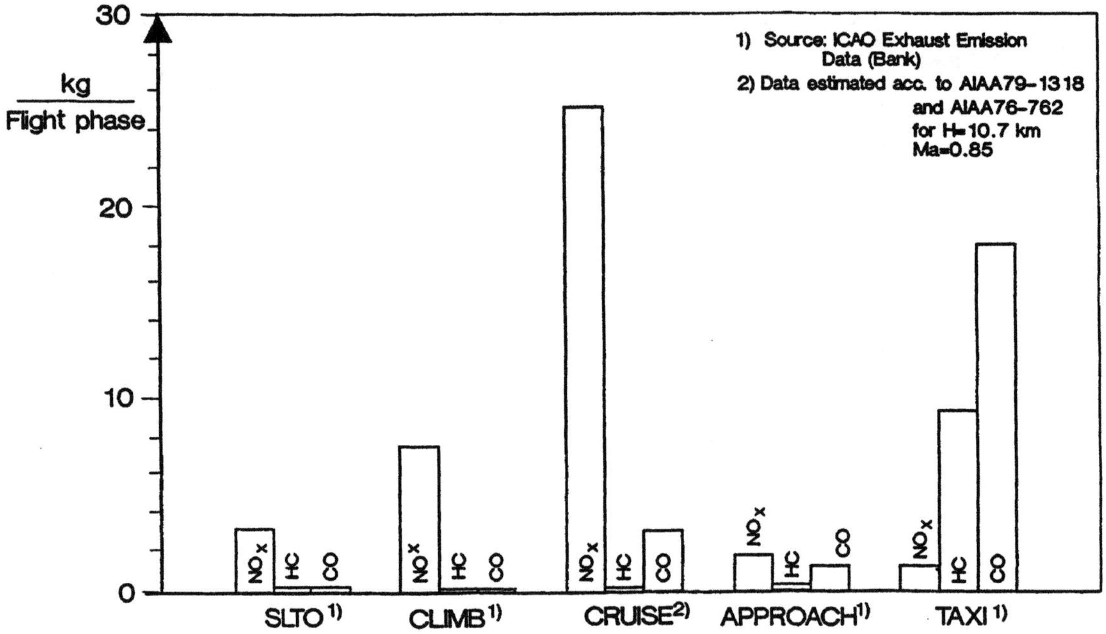

Fig. 9

Particle Emission for Various Phases of Flight
Engine CF6 – 50C

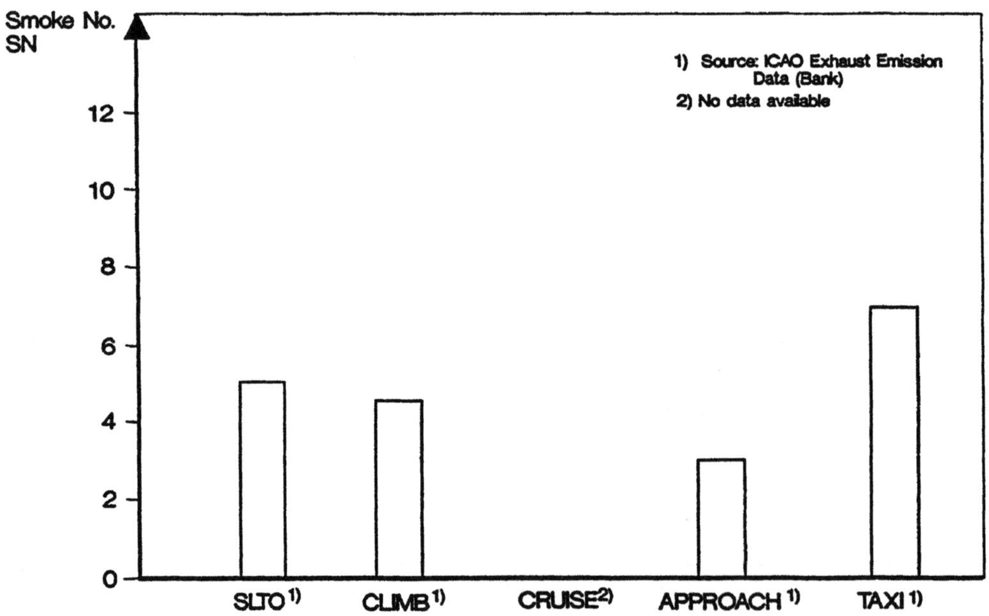

Fig. 10

Shrouded Propfan as Technical/Economical Compromise between Turbofan and Unshrouded Propfan

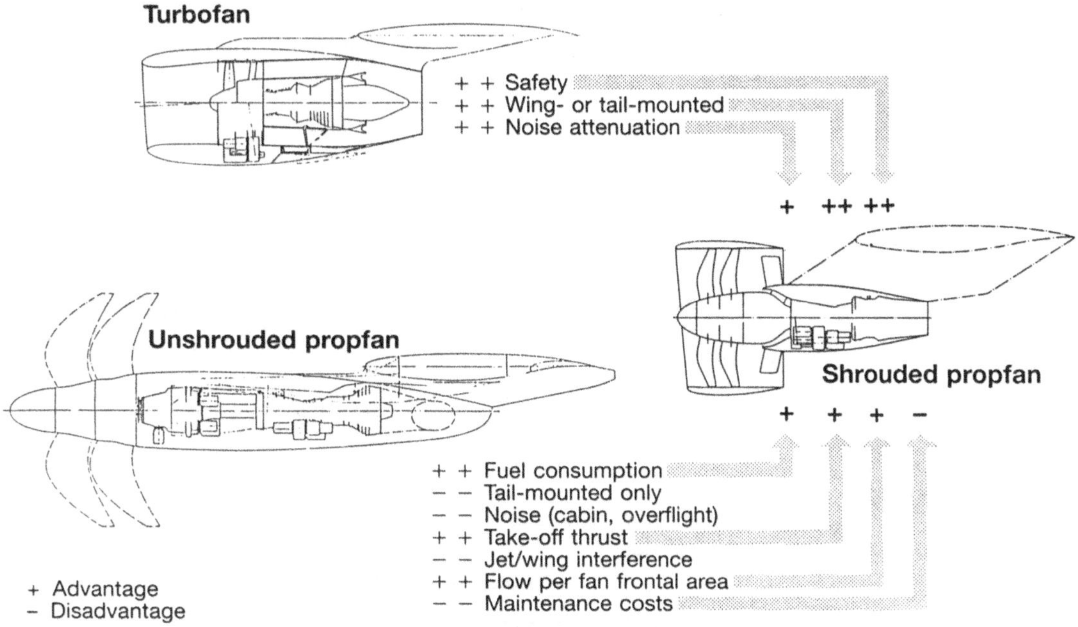

Fig. 11

Shrouded Propfan Concepts

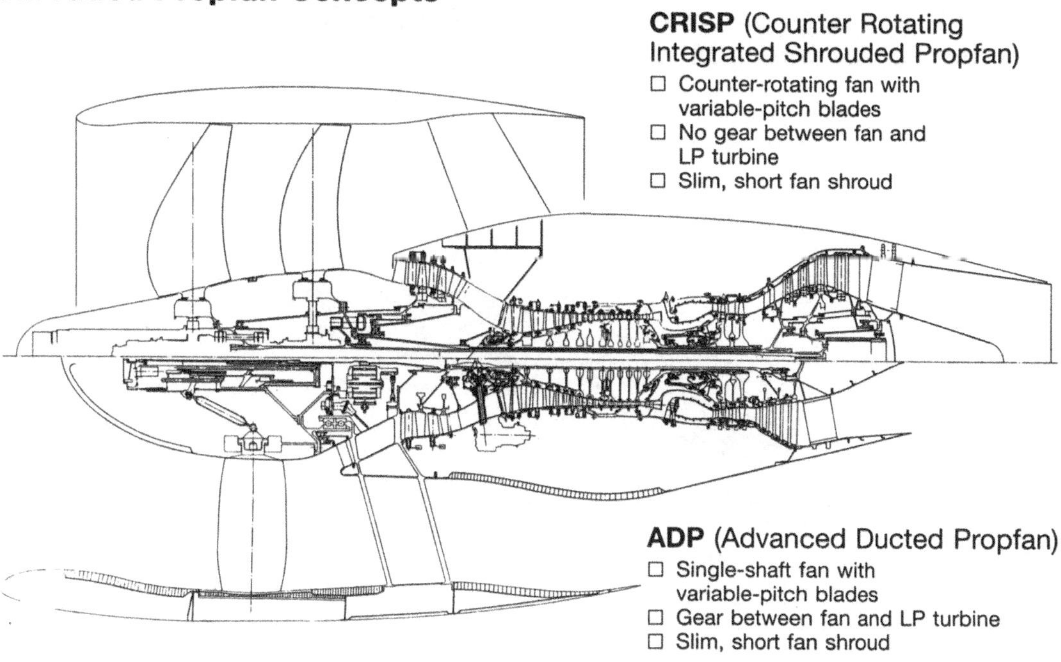

Fig. 12

Development History of Cycle Data of Engines for Commercial Aircraft

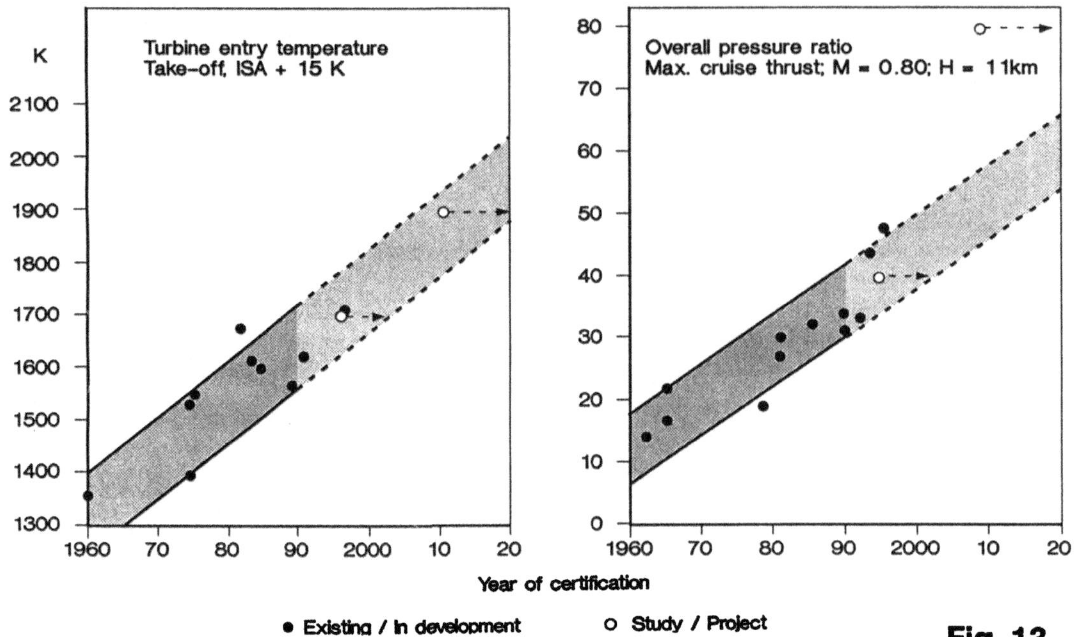

Fig. 13

Comparison between Analytical and Statistical Cycle Data of Planned Engine for Commercial Aircraft

Max. cruise thrust; M=0.80; H=11 km

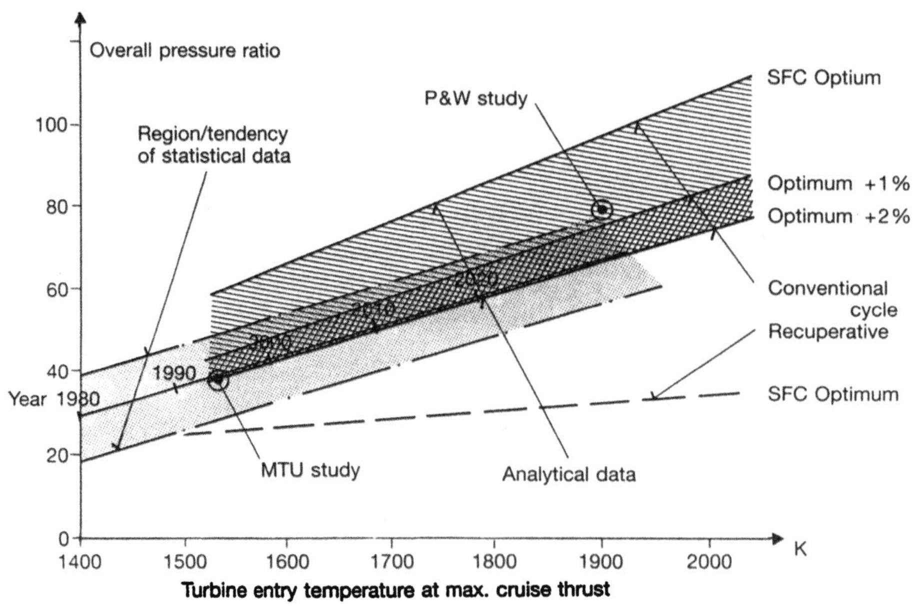

Fig. 14

Development History of Relative Core Engine Mass Flow and Core Engine Thrust Max. cruise thrust; M=0.80; H=11 km

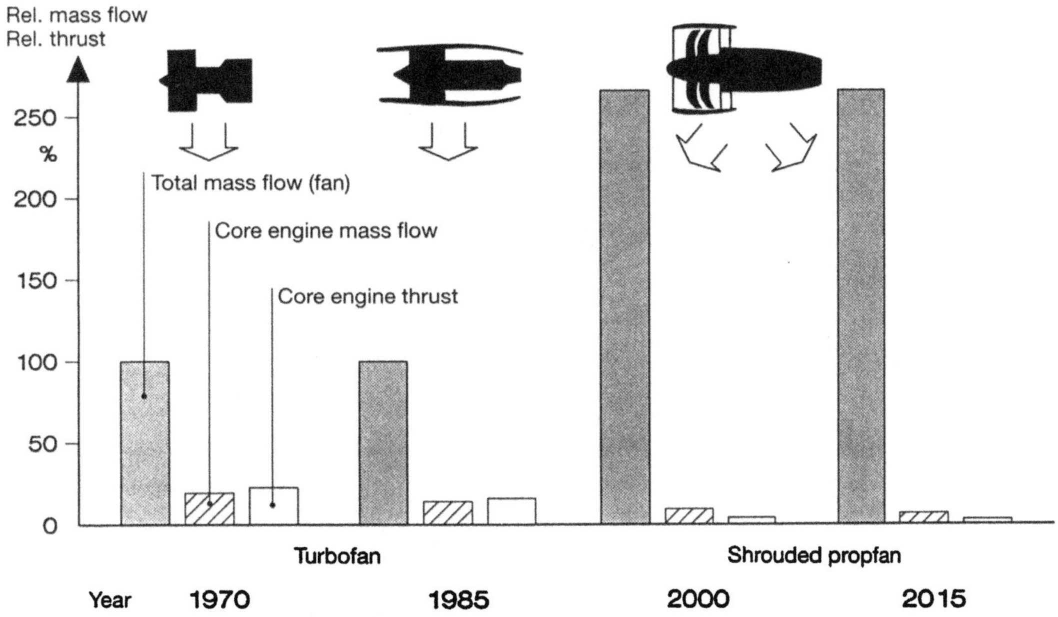

Fig. 15

Future Shrouded Propfan Concepts

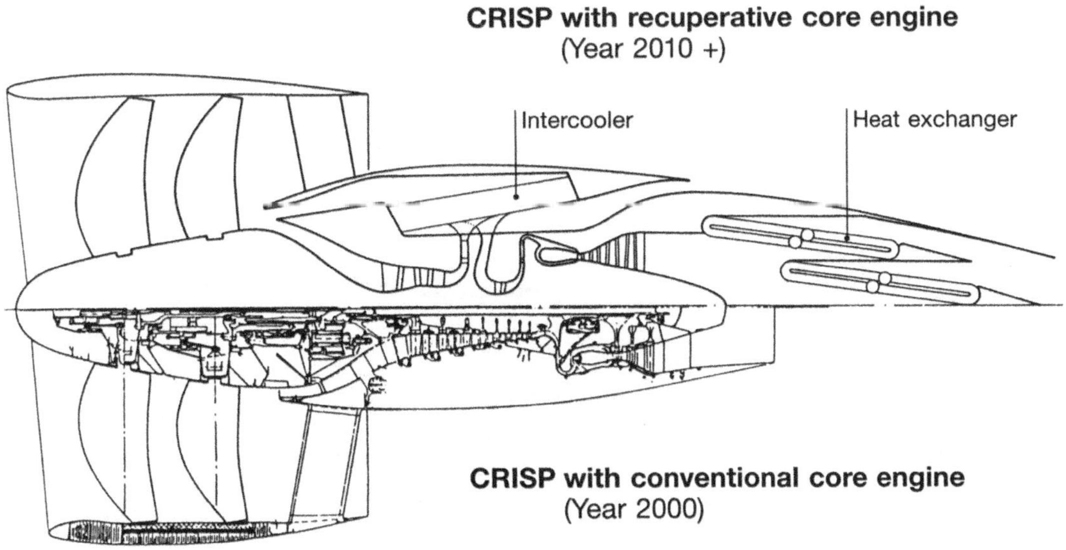

Fig. 16

Specific Fuel Consumption of Planned Shrouded Propfan Engines with Conventional and Recuperative Core Engine Cruise; M=0.80; H=11 km

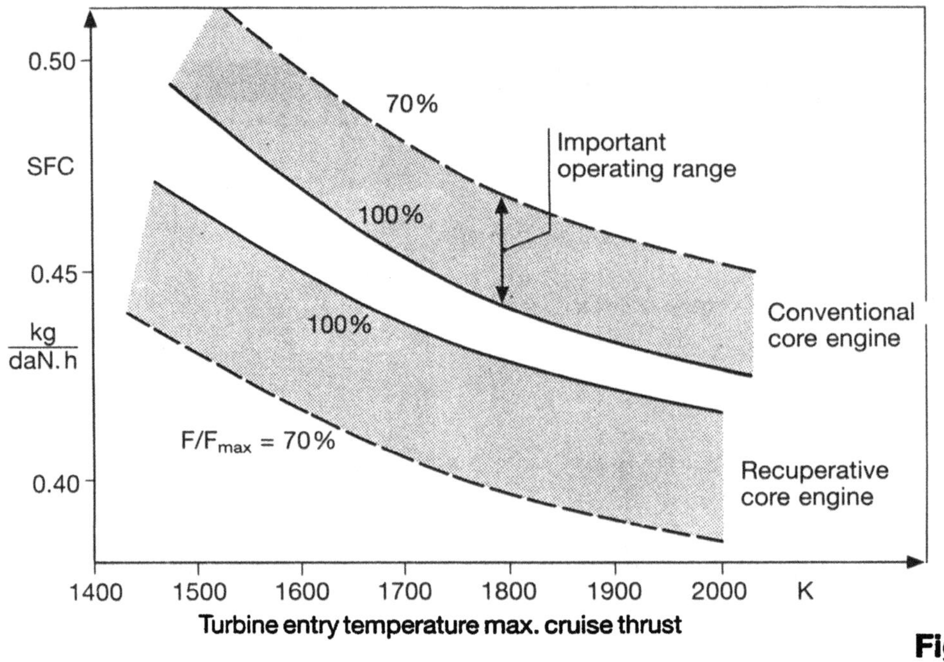

Fig. 17

Expected Improvement in Specific Fuel Consumption of Engines for Commercial Aircraft Max. cruise thrust; M=0.80; H=11 km

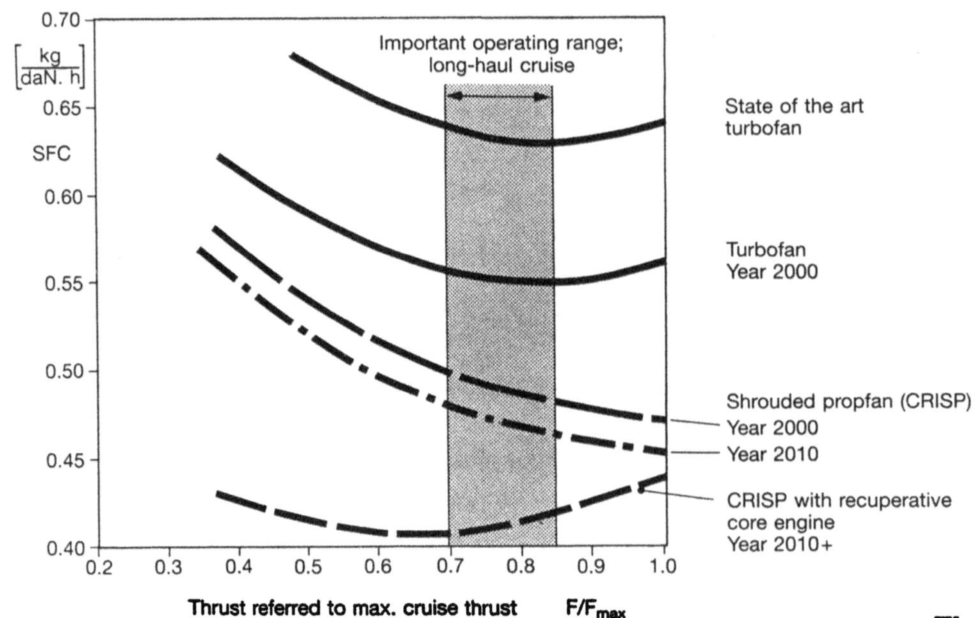

Fig. 18

Estimated Possible NO_x Reduction

Rel. NO_x emission

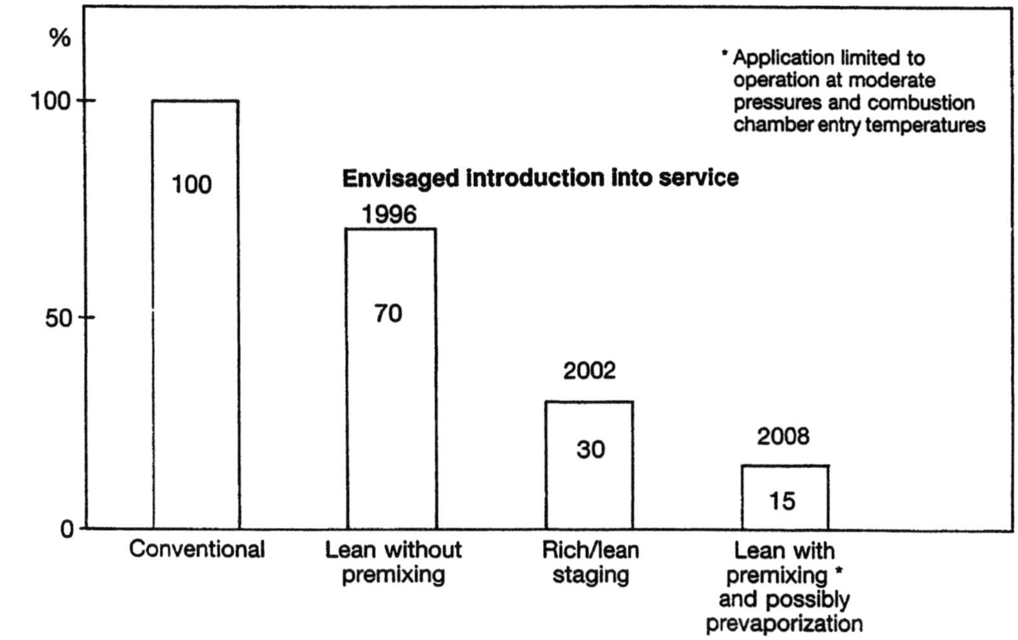

Fig. 19

NO_x Emission Referred to Thrust of Future Shrouded Propfans

Max. cruise thrust; M=0.80; H=11 km

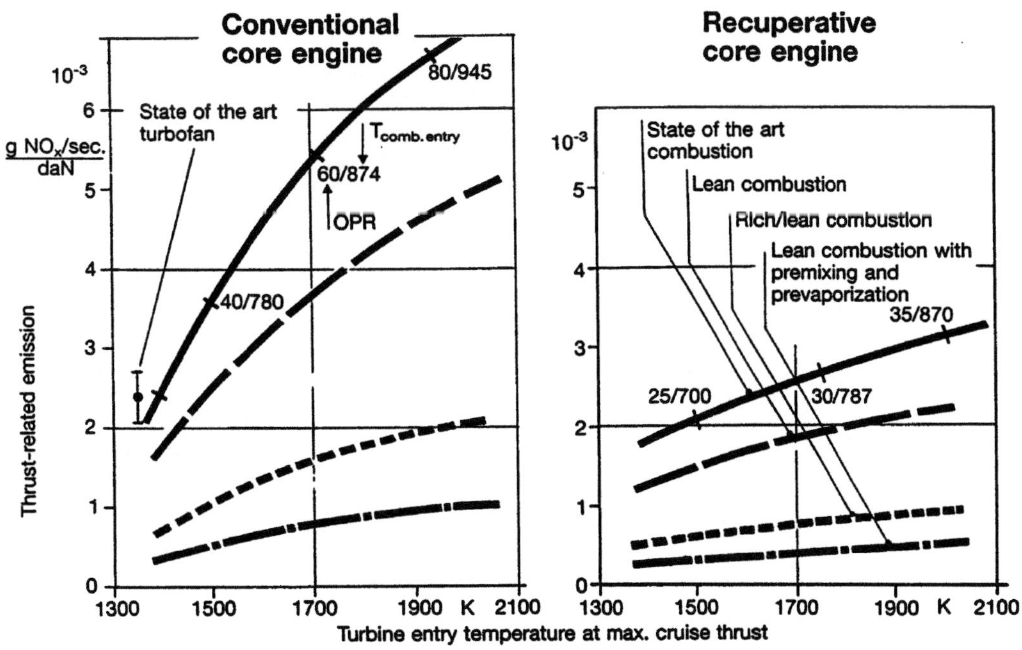

Fig. 20

Principle of a Gasturbine Combustor

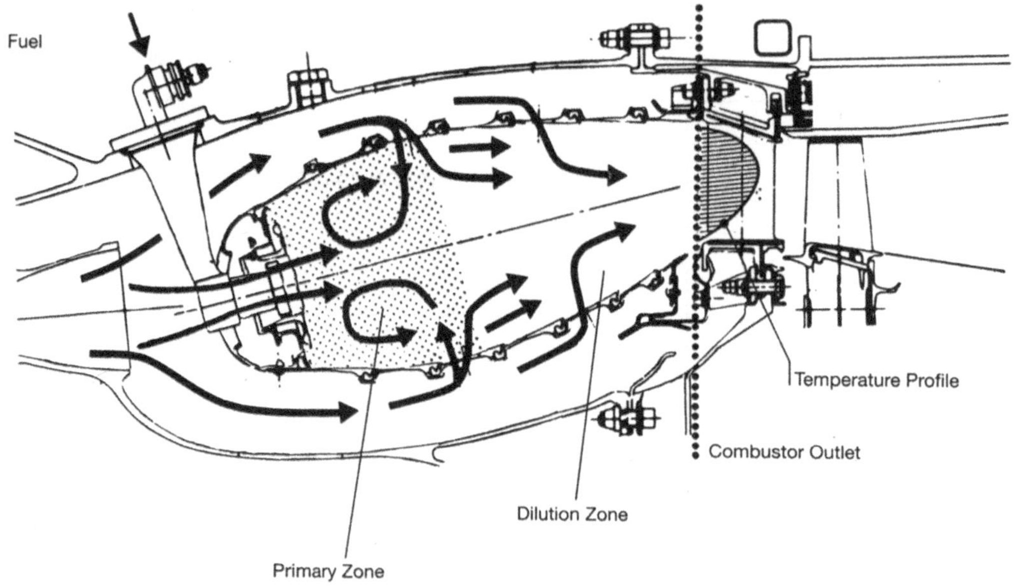

Fig. 21

Conventional Combustor

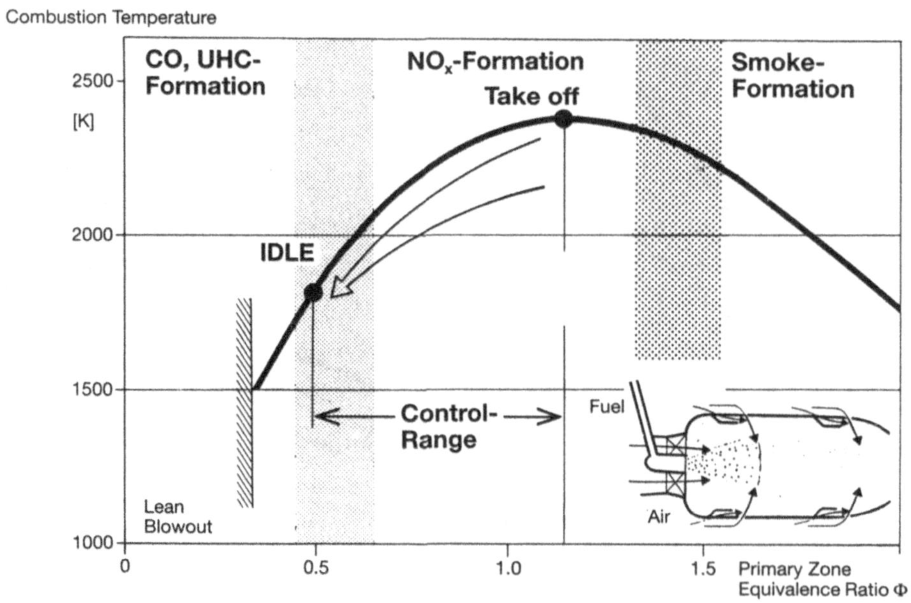

Fig. 22

Lean Combustion

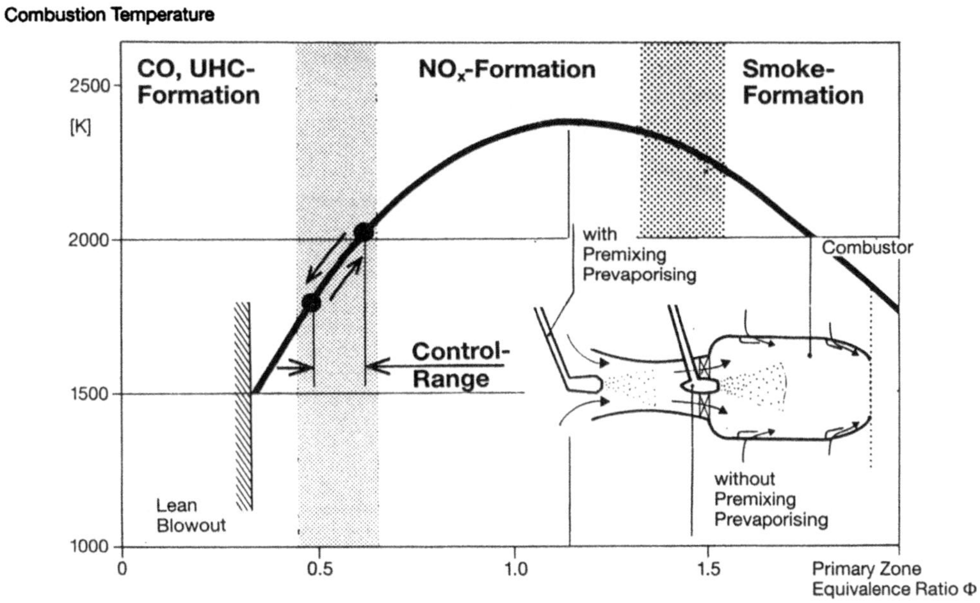

Fig. 23

The Problem of Autoignition in the Premix Duct of a LPP-Combustor

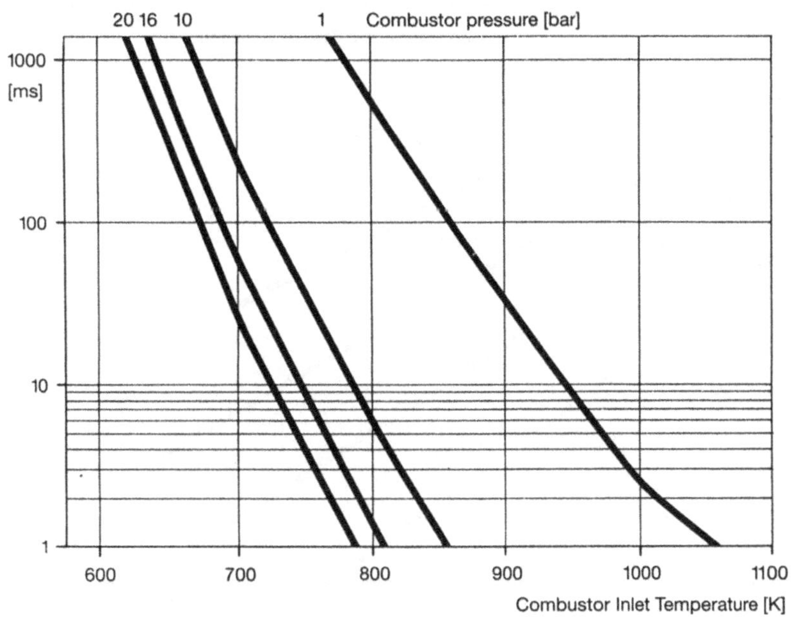

Industrial Gas Turbine
P_{inlet} = 11 bar
T_{inlet} = 600 K
$\tau = 2.76 \cdot 10^4$ ms

Aviation Gas Turbine
P_{inlet} = 23 bar
T_{inlet} = 750 K
$\tau = 6$ ms

Fig. 24

Predicted Influence of Residence Time on the Degree of Vaporisation

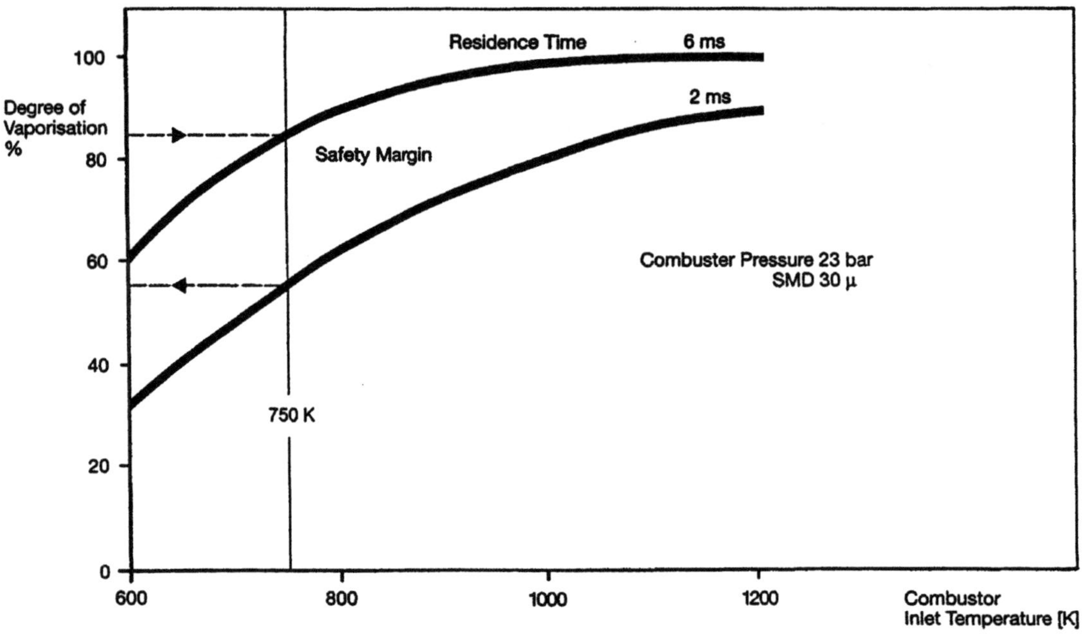

Fig. 25

Influence of Combustor Exit- and Primary Zone-Temperature on the Available Amount of Cooling Air (Dilution Air Not Considered)

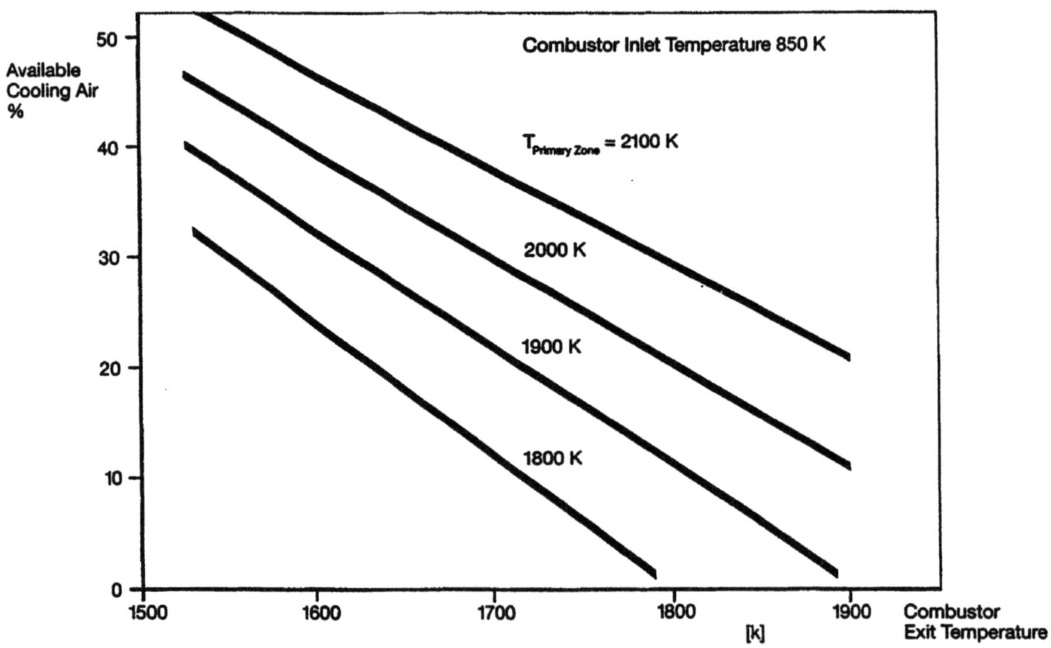

Fig. 26

Rich-Lean Combustion

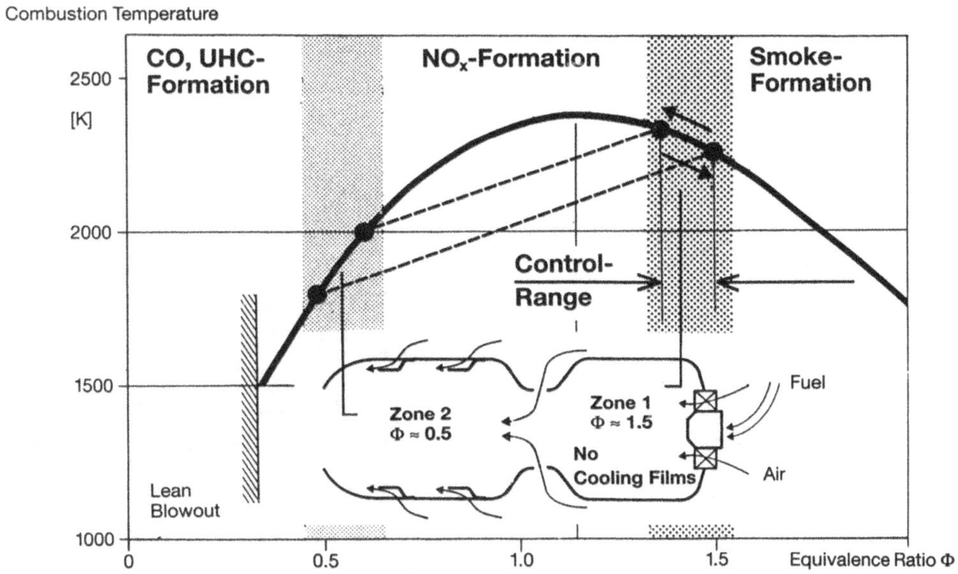

Fig. 27

Predicted Total Fixed Nitrogen as a Function of Primary Zone Equivalence Ratio and Residence Time

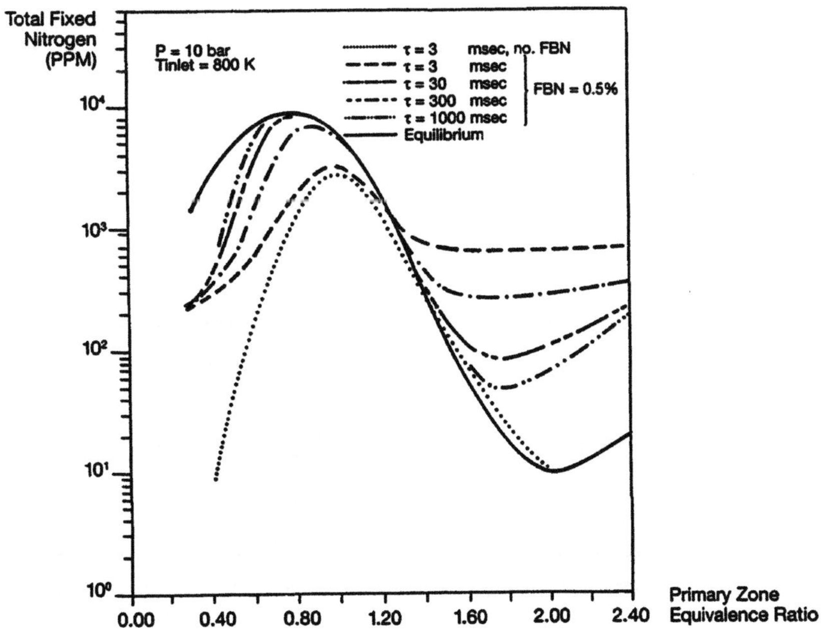

Fig. 28

Different Methods of Primary Zone Fuel-Air Control

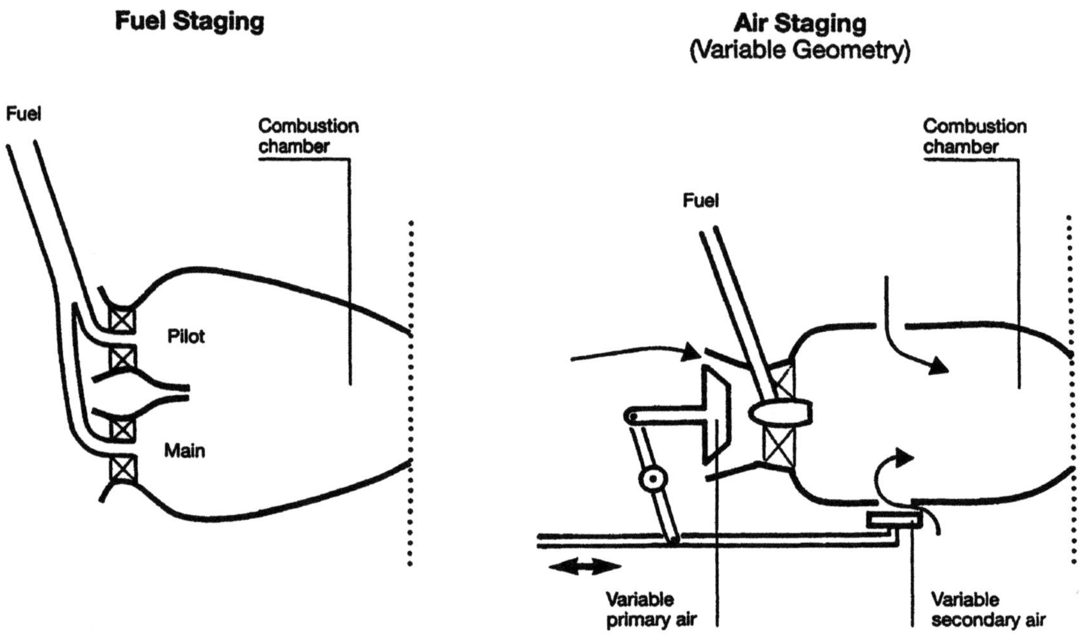

Fig. 29

Aviation Fuels

Comparison between mass and volume for equivalent energy-content

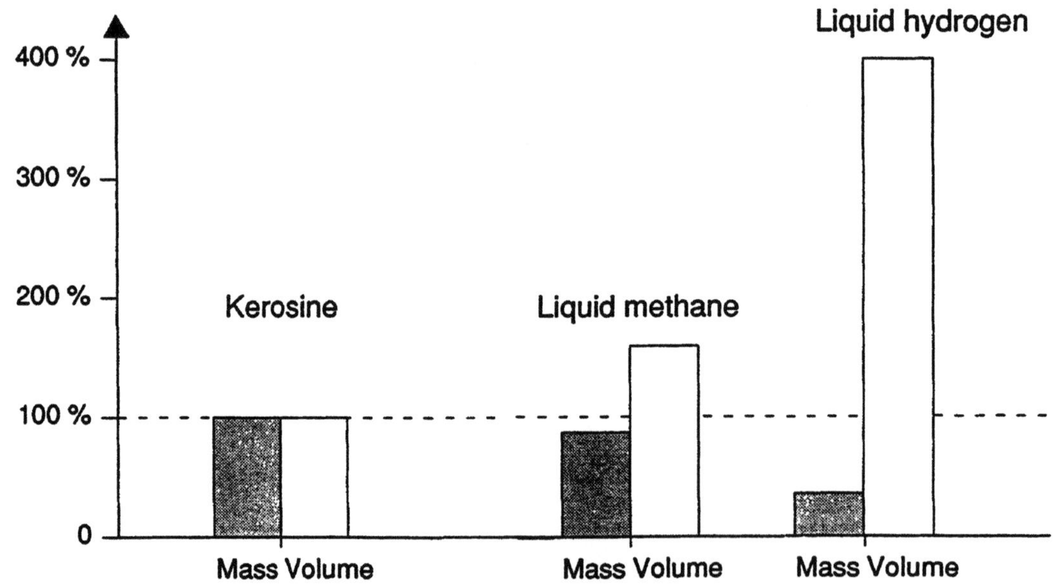

Fig. 30

Potential for Reduced NO_x-Emission Combustion of Different Fuels

	Ideal fuel	Kerosine	Methane	H_2
Spontaneous ignition	High temperatures and pressures without spontaneous ignition	–	+	+
Flashback	Low flashback velocity	+	+	–
Mixing rate	High	–	+	+
Combustion limits	Narrow	+	+	–
Stoichiometric temperature	Low	+	+	–

+ Favourable
– Unfavourable

Fig. 31

Ignition Limits of Hydrogen, Methane and Kerosine in Air
(Spark ignition, gaseous, T = 300 K)

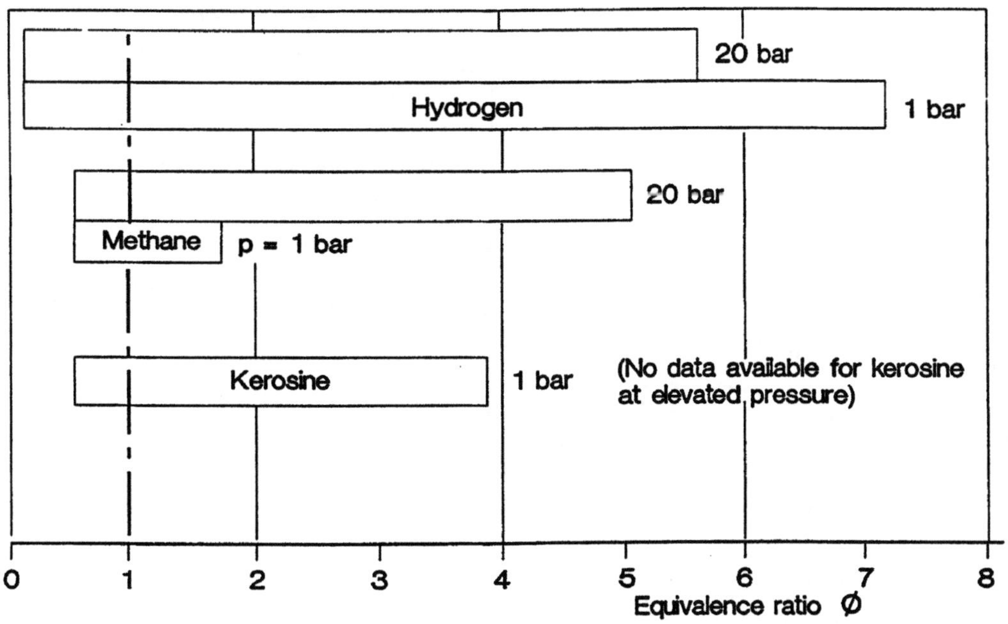

Fig. 32

Experimental Investigations into NO_X Formation for Combustion of Hydrogen

Model combustion chamber

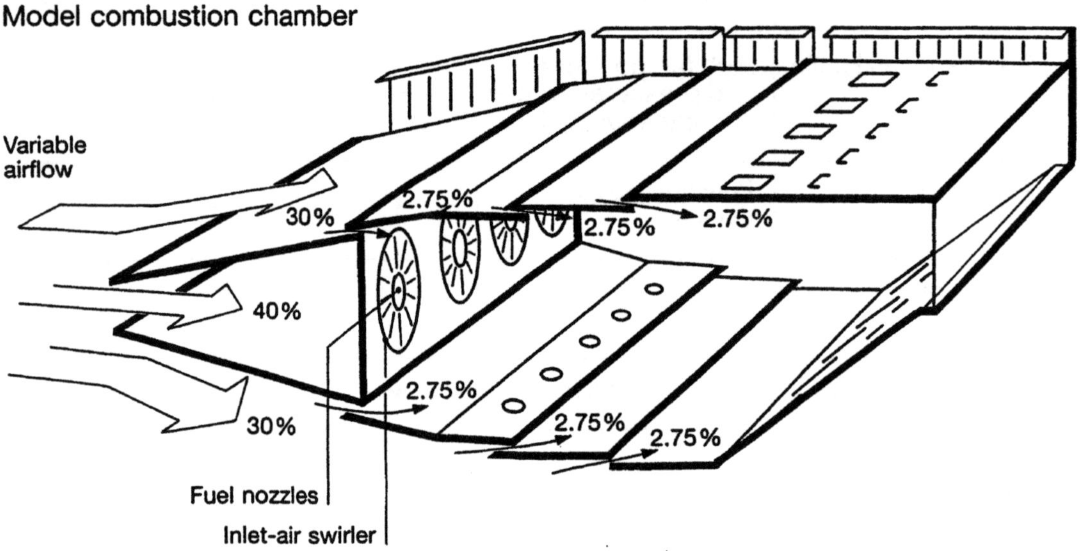

From D.N. Anderson: Emissions of Oxides of Nitrogen from an Experimental Premixed Hydrogen Burner
(Ref 6) NASA TM X-3393, 1976

Fig. 33

Comparison of NO_X Emissions for Combustion of Kerosine and Hydrogen

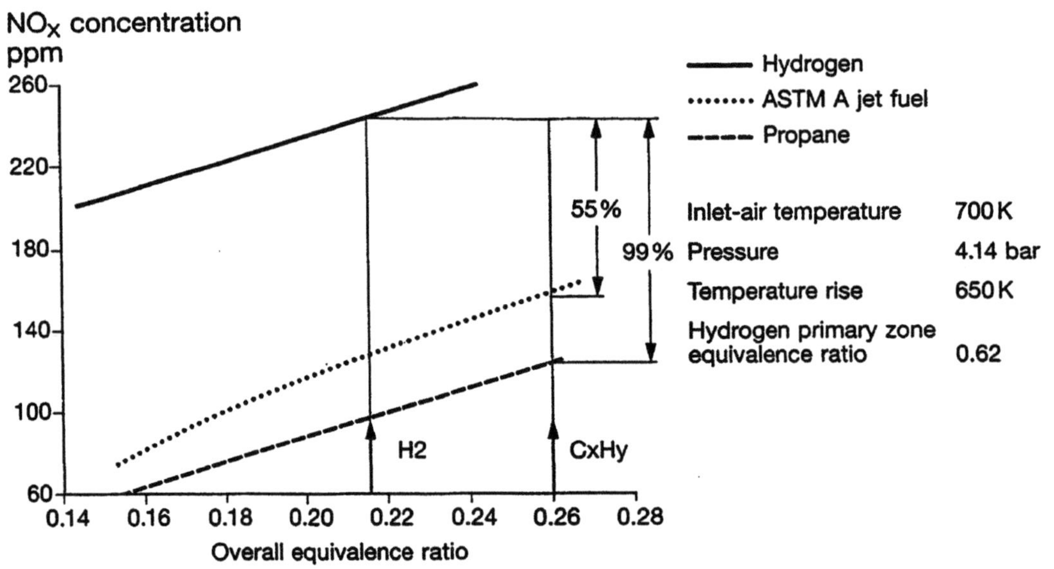

From D.N. Anderson: Emissions of Oxides of Nitrogen from an Experimental Premixed Hydrogen Burner
(Ref 6) NASA TM X-3393, 1976

Fig. 34

Emissions with Combustion of Hydrogen, Kerosine and Methane

Basis: 10 MJ fuel (corresponding to 1.2 l liquid hydrogen or 0.3 l Kerosine)

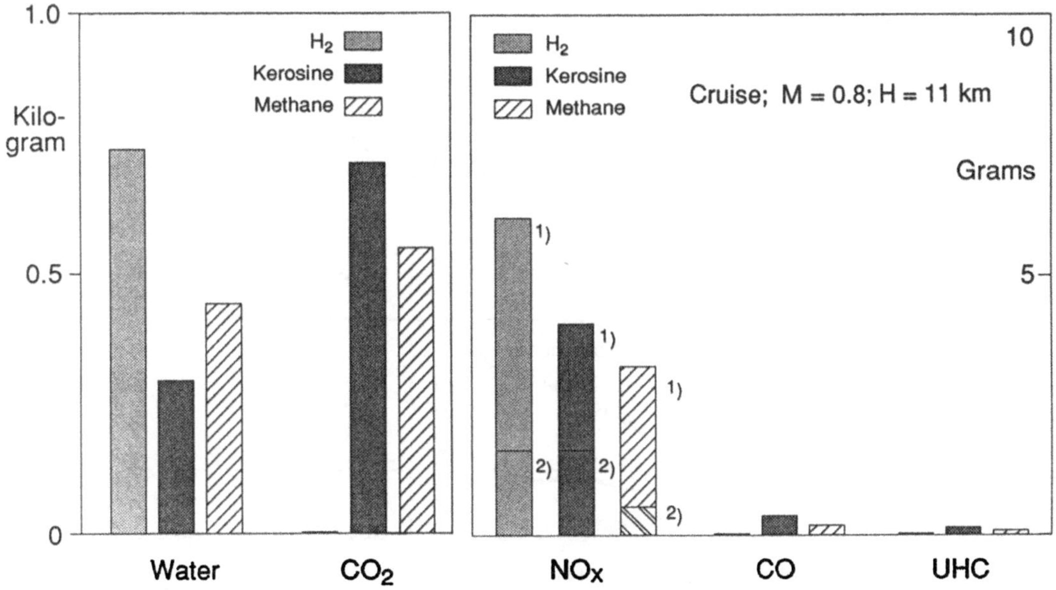

1) With conventional combustion chamber
2) With reduced emission combustion chamber

Fig. 35

Aviation Fuels

	Jet A	Liquid methane	Liquid hydrogen
Cooling capacity	−	+	+
Tank	+	−	− −
Control	+ +	−	− −
Safety	+	−	− −
Environmental acceptability	+	+ +	−
Price	+	−	− −
Infrastructure	+ +	−	− −

+ + Highly favourable
− − Highly unfavourable

Fig. 36

Development of air traffic worldwide

Airline and charter operations

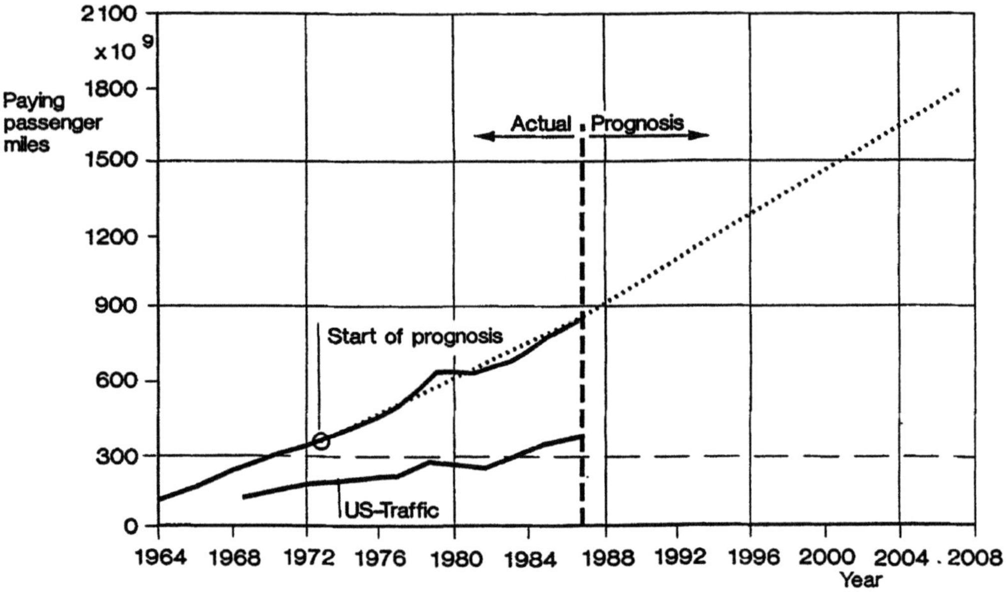

Fig. 37

Potential for Reductions in Fuel Consumption per Passenger Kilometer of Subsonic Aircraft by Technical Progress in the Time Frame 1990–2010

*Secondary Effects on Aircraft Weight Not Considered

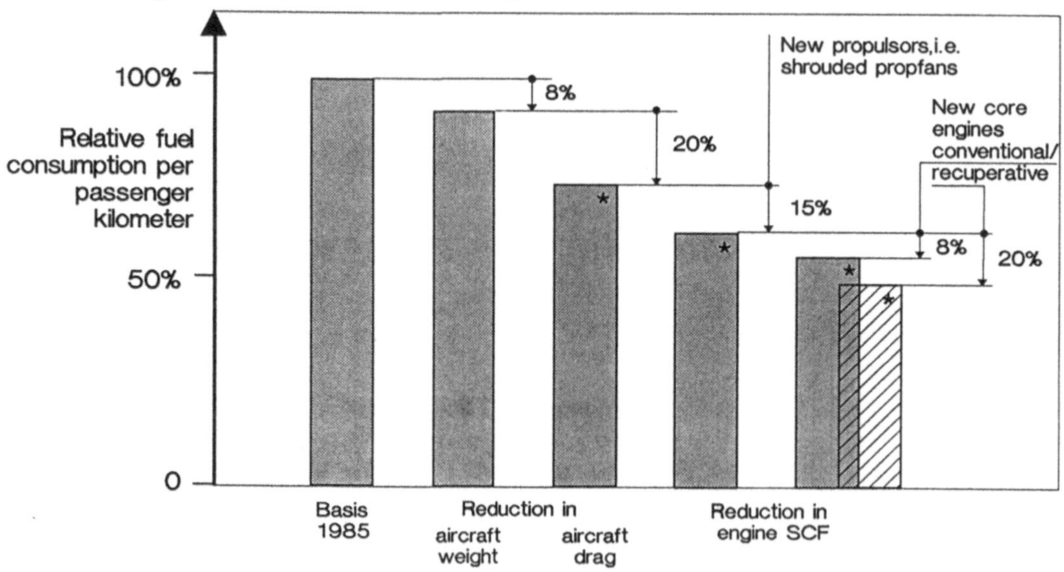

Fig. 38

ASSUMED TECHNICAL STANDARDS AND SERVICE INTRODUCTION TIMES OF FUTURE AIRCRAFT/ENGINE GENERATIONS

Standard	Service Introduction	Aircraft			Engine		
		Drag %	Weight %	Thrust per PAX %	Concept	SFC %	EI-NO_x **** %
0	-	-	-	105	Average of all engines in service	110	100
A	1986	100* (98)**	100	100 (98)	3rd generation turbofan; convent. combustor	100 (94)	100 (100)
B	1996	95 (93)	96	91 (89)	UHB turbofan lean combustor	90 (84)	73 (73)
C	2006	90 (88)	92	83 (81)	Shrouded propfan with conventional core engine; rich/lean combustor	80 (74)	33 (33)
D	2016	80 (78)	90	74 (72)	Shrouded propfan with ICR core engine; lean/premix.combustor	68 (63)	9 (9)
(E)***	2026	72 (70)	90	65 (63)	Not defined (fictitious)	57 (53)	6 (6)

* Open numbers are valid at service introduction
** Numbers in brackets are valid after 16 years of service
*** For extrapolation only
**** Related to thrust, standard A, 1986 ≙ 100%

Fig. 39

Progress in Replacement of Generation A by Generation B
(Aircraft and/or Engine)

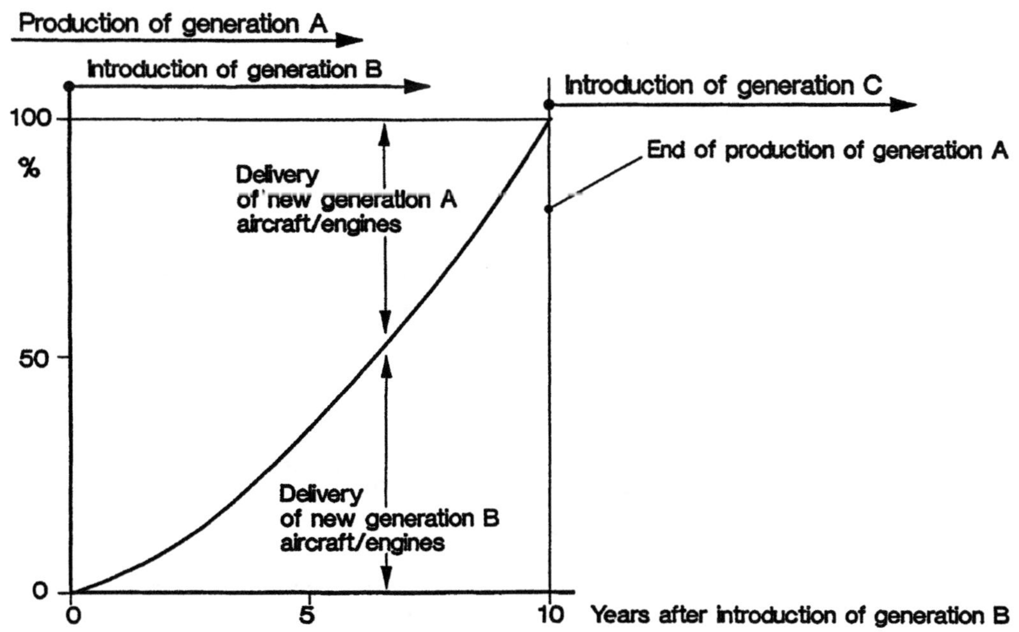

Fig. 40

Rate of Attrition and Phasing Out of an Aircraft/Engine Family Referred to Total Number of Aircraft Supplied
(Simplified)

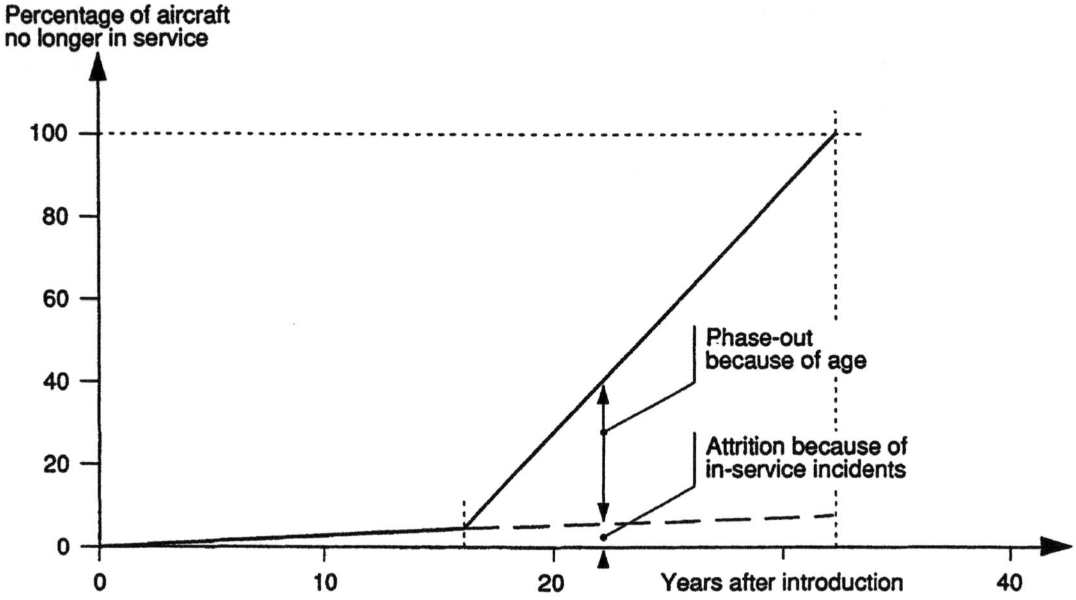

Fig. 41

Influence of Successive Introduction of Advanced Aircraft/Engine Generations on H_2O, CO_2 and NO_X Emissions

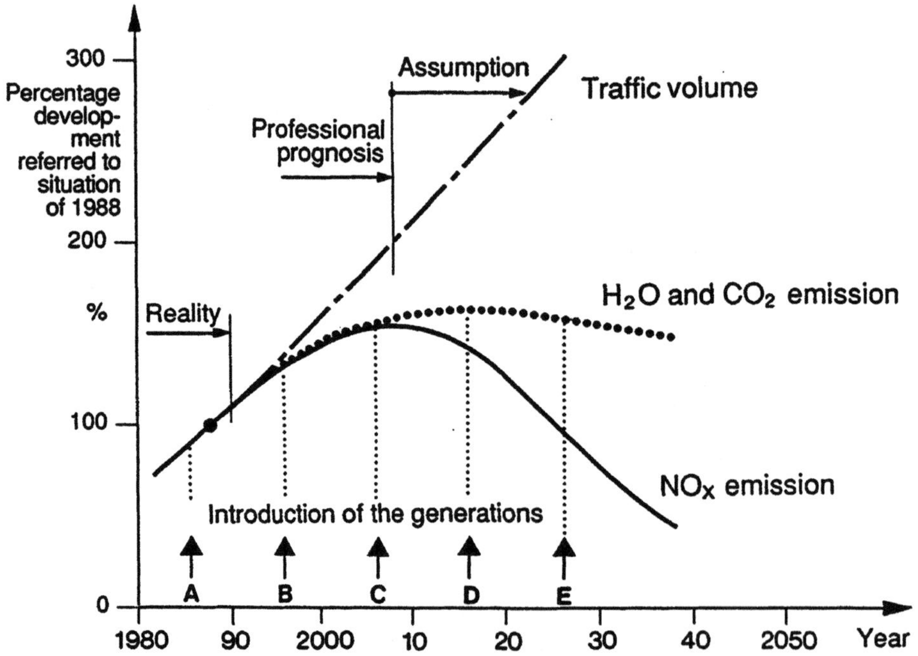

Fig. 42

Oil Supply Capacities as of 1988

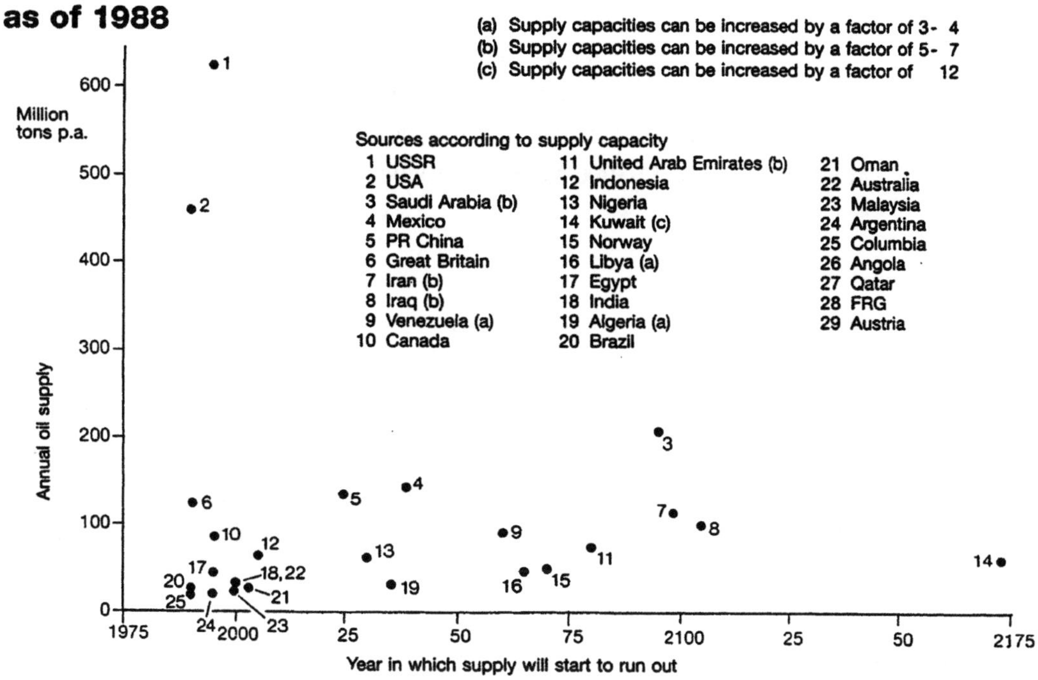

Fig. 43

Marginal Oil Producer Prices for New Development Projects

Marginal oil producer prices in US-$ per barrel

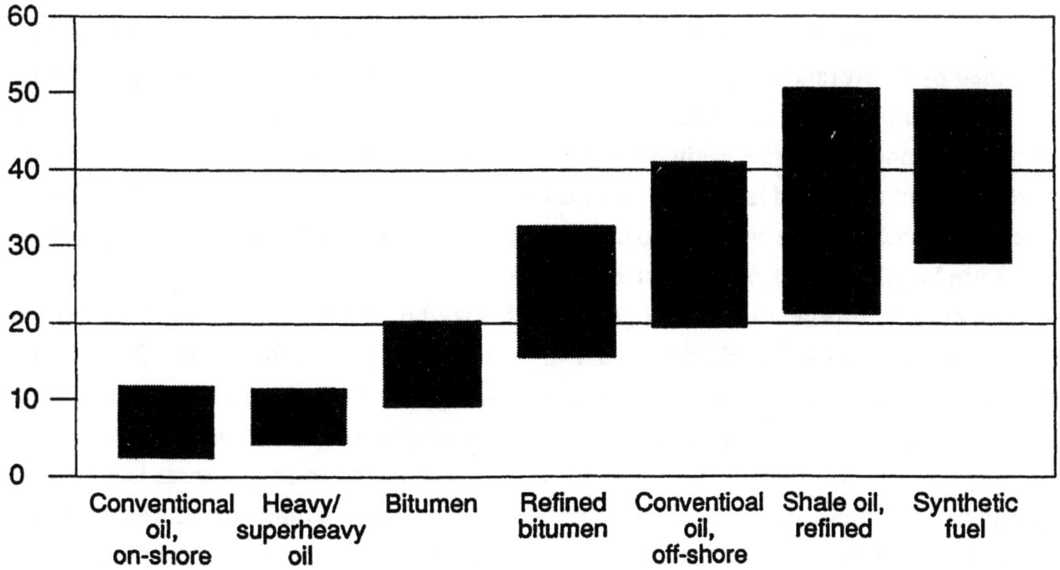

Fig. 44

CONSTITUENTS AND TRANSPORT PROPERTIES OF THE ATMOSPHERE ABOVE AND BELOW THE TROPOPAUSE

Peter Fabian
Ludwig-Maximilians-Universität München
Lehrstuhl für Bioklimatologie und Angewandte Meteorologie
Amalienstraße 52, D-8000 München 40, Fed.Rep. of Germany

ABSTRACT

Dynamical and photochemical properties of the troposphere and stratosphere are briefly discussed. The impact of aircraft exhaust contaminants is assessed. Due to long residence times at cruise altitude, some exhaust products, in particular H_2O and NO_x, are likely to build up considerable enhancements of natural background mixing ratios, with possible consequences on ozone and climate.

INTRODUCTION

The Earth's atmosphere constitutes a very thin gas layer. Compared to the dimension of the solid planet, its vertical extension is almost negligible: more than 90 % of its mass are concentrated in the lowest 10 to 15 km , the troposphere, where the weather processes take place. At 100 km altitude, the air pressure has already dropped to 10^{-6} of its surface value. If this height would be defined as the upper boundary of the atmosphere, it would account for less than one percent of the Earth's diameter.

Nevertheless, despite of this small extension, the atmosphere has a characteristic vertical structure determined by its average temperature profile, which is shown in figure 1 (lower part). The lowest layer or troposphere is characterized by temperatures decreasing with height. Above the tropopause, temperatures rise again up to the stratopause at about 50 km, where temperatures are almost as high as at ground level. This layer called stratosphere acts as a huge inversion. Compared to the troposphere where vertical mixing by turbulence and convection is pretty fast, vertical motions are largely suppressed in the stratosphere.

The temperature profile shown in figure 1 reflects the thermal processes activated by solar radiation, whose spectral distribution is shown in the upper part of the figure. The visible part of the solar spectrum penetrates down to ground level, partly attenuated by gases and aerosols. It heats the Earth's surface, which in turn heats the atmosphere. Thus temperatures decrease with increasing distance from this surface. The ultraviolet part of the solar spectrum is absorbed in the stratosphere and mesosphere by oxygen and ozone thereby creating the aforementioned warm layer, which is unique in the whole solar system as no other planet possesses appreciable amounts of oxygen and thus ozone. Above 100 km, another temperature increase marks the thermosphere, where absorption of short ware solar UV and X rays cause temperatures to increase up to 1500 degrees.

With respect to chemical and dynamical aspects, stratosphere and troposphere are extremely different. By themselves, atmospheric constituents would not react with each other at all. In the stratosphere (and mesosphere as well), however, UV radiation photolyses most of the atmospheric gases thereby liberating atoms and radicals, which in turn drive a complex photochemical reaction scheme. One of the most important complexes of this scheme is the formation of the ozone layer.

No solar UV with wavelengths shorter than 290 nm can reach the troposphere. Thus except for ozone and pollutants such as NO_2, almost no constituent can be photolysed in this lowest part of the atmosphere. Nevertheless, a complex photochemical reaction system governs the troposphere, too. It is driven by the hydroxyl radical OH, a product of a two-step reaction initiated by ozone, which is permanently mixed down from the stratosphere:

$$O_3 + h\nu(\lambda < 310\,nm) \longrightarrow 0(^{\prime}D) + O_2$$
$$O(^{\prime}D) + H_2O \longrightarrow 2\,OH$$

OH oxidises many of the trace gases entering the atmosphere by both natural and anthropogenic processes, thereby forming soluble substances, which are washed out by precipitation.

TRANSPORT PROPERTIES OF THE TROPOSPHERE AND STRATOSPHERE.

In the troposphere, turbulence and convection provide effective vertical mixing. If an inert tracer is injected, it is rapidly carried along with the winds and thus distributed within the respective hemisphere. This "hemispheric mixing time" is of the order of one to two months. It takes considerably longer until global mixing of a tracer is achieved. Since the Intertropical Convergence Zone (ITCZ) acts as a barrier, the "interhemispheric mixing time" is of the order of one to two years.

Weather processes in connection with the water cycle of evaporation, cloud formation and precipitation are the most important cleaning mechanism of the atmosphere. Particles and soluble gases

are directly involved in cloud formation. They are washed out by precipitation and thus have a short tropospheric residence time, which is of the order of days to weeks. The residence times of non-soluble gases which are converted into soluble species depend upon the rate this conversion proceeds. Pollutants such as NO_2 are oxidised by OH to HNO_3 which is washed out by rain. The residence time of NO_2 is about 3 days which is much shorter than both the hemispheric and the interhemispheric mixing time. Thus NO_2 (and its precursor NO as well) is not distributed very far from its emission sources. Like SO_2 it pollutes its regional environment only. Constituents, whose residence times are much longer than the interhemispheric mixing time, are almost equally distributed over the whole globe, independent of the location of the sources. N_2O, with a residence time of about 100 years, belongs to this category. Fig 2 shows atmospheric

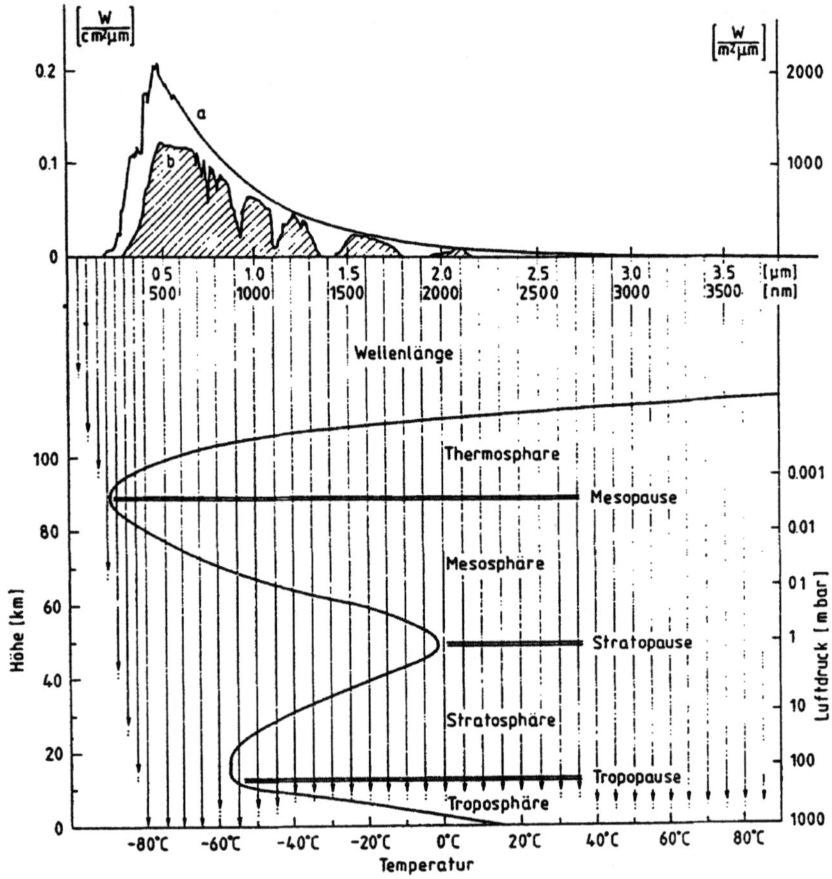

Figure 1: Spectral distribution of solar radiation (upper part),a outside the Earth's atmosphere, b at ground level.Temperature distribution and layer structure of the atmosphere (lower part). The warm layer characterized by a temperature maximum at the stratopause is a result of heating due to absorption of solar radiation by ozone. The vertical arrows indicate schematically how deep solar radiation of the respective wave length penetrates (after Fabian 1989).

residence times for various natural and man-made constituents ranging from seconds to a century.

Because of its vertical temperature structure, the stratosphere acts as a huge inversion thus suppressing vertical mixing. Vertical motion is largely maintained by quasi-horizontal motion on isentropic surfaces. It is driven by large-scale waves and eddies and results in poleward transport connected with sinking motion. This quasi-horizontal transport is responsible for the distribution of ozone or any other substance in the stratosphere.

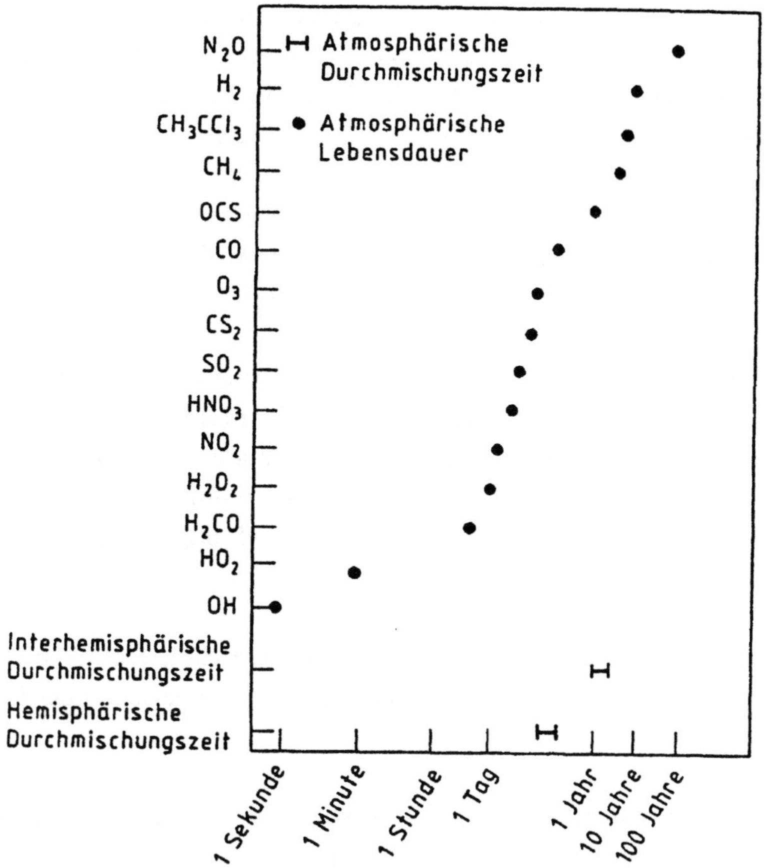

Figure 2: Atmospheric residence times of trace gases vary between one second and one century (after Chameides and Davis 1982).

Stratospheric transport processes have been studied by means of radioactive debris from high yield nuclear explosions performed in the atmosphere. Between 1967 and 1971 such atmospheric nuclear tests were caried out by Red China (at 40° N) and by France (at 22° S). Approximately every 3 months fission products were sampled by high-flying aircraft and balloons between 75° N and 55° S within "Project Airstream" by the Health and Safety Laboratory, US Atomic Energy Commission. Thus two-dimensional hight-latitude distributions of various nuclides were obtained. A total of 18 two-dimensional decay-corrected distributions of Zr-95 (65 days half-life) were provided by K. Telegadas (1973) for 2-D model studies (FABIAN 1974).

Fifteen of these decay-corrected Zr-95 cross sections were suitable for use as initial tracer distribution. By integrating the transport equation, theoretical cross-sections were obtained for the high-altitude sampling periods thereafter and compared with the experimental data. Every experiment was run until the distribution of debris was disturbed by another nuclear blast. Figure 3 shows three of the fifteen experiments with the Zr-95 originating from the 1970 - 71 nuclear explosions on both hemispheres.

Isolines of the Zr-95 mixing ratio (units: pCi/standard cubic metre) are plotted on a latitude-altitude grid, the theoretical predictions from the model being solid lines for comparison with the high-altitude sampling data (dashed lines).

For comparison the altitudes of highest concentration, symbolized by crosses and dots for theoretical and experimental data, respectively, are also given.

During the sampling period 22 - 27 February, 1971, debris from the French 31 May, 4 July (southern hemisphere) and the Chinese 14 October, 1970 (northern hemisphere), nuclear tests was collected. This material was taken as initial distribution for the first experiment. Figures A and B give the comparison between model prediction and sampling periods, 22 - 28 May and 19 - 23 July, 1971, respectively.

Based on the initial distribution from the 22 - 28 May, 1971, sampling, Figs. 3 C and D give results for the 19 - 23 July and 4 - 6 October, 1971, periods, respectively

Debris from another French explosion on 13 June, 1971 (1 MT), together with the residue of the Chinese 14 October, 1970, shot dominating during the 19 - 23 July, 1971, sampling period, was taken as the initial distribution for the last experiment. Figure 3 E gives predicted and measured cross-sections for the last sampling period of this series 4 - 6 October, 1971.

Data, classified as mixture, from different nuclear explosions according to their Zr-95/Cs-144 ratio, were not included in these model experiments. Thus some of the dotted lines show gaps, especially in the equatorial region where the mixing between debris from the French and Chinese bombs mostly occurred. A detailed presentation of the model is given elsewhere (Fabian 1974).

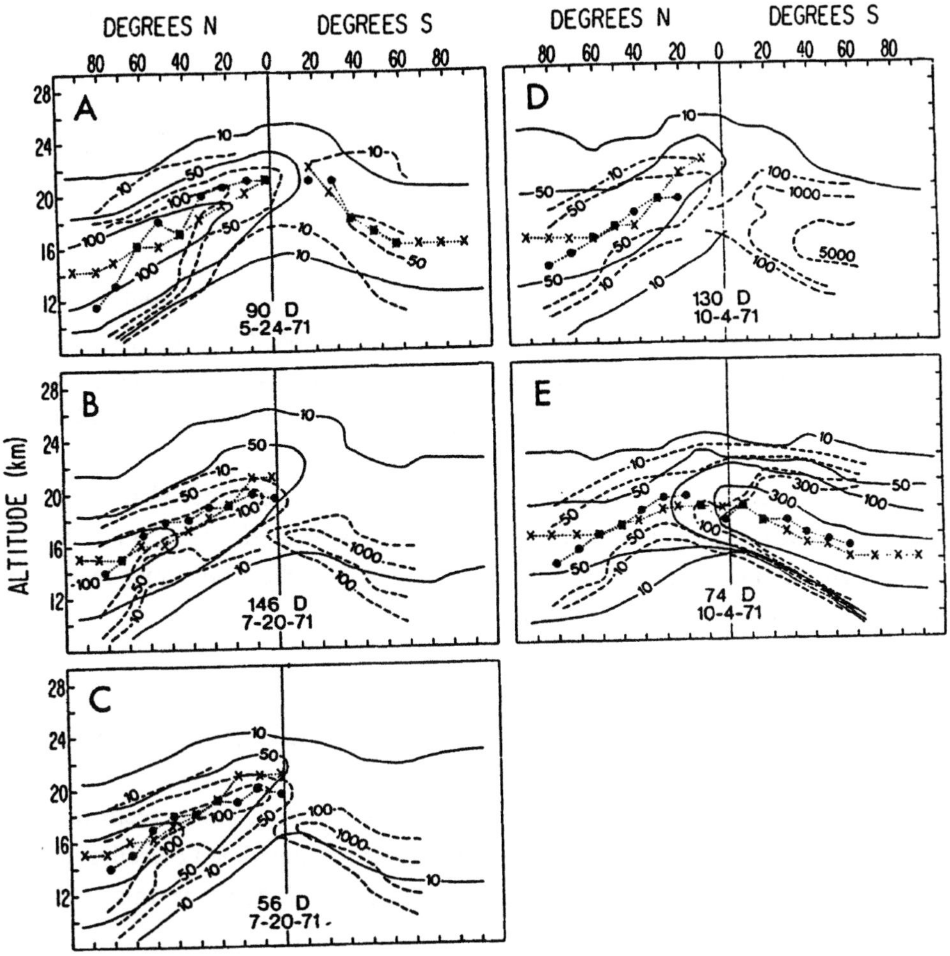

Figure 3: 31 May, 4 July (S), 14 October, 1970 (N), and 13 June, 1971 (S), nuclear tests. Distribution of Zr-95 mixing ratio (p Ci/SCM) corrected for radioactive decay to 14 October, 1970. Solid lines: theoretical prediction from model; dahed lines: data from high-altitude sampling program. Altitudes of maximal concentration are indicated by x and . for theoretical and experimental data, respectively. Experiment 1 (sub-figures A and B) based upon the initial distribution from the 22-27 February 1971, sampling (4 und 8 months after injection, respectively). Experiment 2 (sub figures C and D) based on the initial distribution from the 22-28 May, 1971, sampling (7 and 11 months after injection, respectively). Experiment 3 (sub-figure E) based on the initial distribution from the 19-23 July, 1971, sampling (9 and 1 month after injection, respectively).The number of days and the date given in every sub-figure corresponds to the integration time and the centre of the sampling interval for which comparison between prediction and observation is made, respectively. For further information see text

Additional experiments were carried out with the tracer model for estimating stratospheric residence times as a function of altitude and latitude. Applying the concept of residence times one has to be very careful because they strongly depend upon how they are defined. Here we were interested in estimating the time for the e-fold decrease of a stratospheric disturbance caused by a single injection at a certain altitude and latitude, for instance by a nuclear bomb explosion or aircraft exhaust products.

Figure 4 demonstrates the results as isolines of residence time derived from the numbers obtained for the grid-points. According to the definition used for this study the residence time for a certain grid-point is that time necessary for reducing the stratospheric disturbance, caused by a single injection at this particular point, to 1/e of its initial value. (For more details see Fabian 1974).

From such results it follows that stratospheric tracers which are removed from the atmosphere by tropospheric scavenging, washout or rainout, have a stratospheric residence time of one year and longer. Residence times increase with increasing height. For the tropopause level a residence time of about 1 year or slightly less can be assumed.

THE IMPACT OF AIRCRAFT EXHAUST ON ATMOSPHERIC COMPOSITION

Based on the fact that aircraft contaminants make up only about one percent of total emissions of contaminants into the atmosphere, the effect of aircraft has long been considered negligible. In the light of long atmospheric residence times at tropopause level, however, this relation by mass only needs to be revised.

Assuming that the bulk of the fuel is burnt during cruise at 10 km altitude, a simple study shows that at least two of the exhaust products, i.e. H_2O and NO_x, cause significant disturbances of the trace gas composition of this altitude regime (Fabian 1988).

For this study it was assumed that worldwide 86 x 10^6 tons of kerosene are consumed per year. For every of the reaction products an emission index was taken from Oliver et al. (1977) according to the following table

Product	Emission Index (g/kg)	Global Emission (t/y)	Increase of mixing ratio at 200 hPa
CO_2	3.22 x 10^3	277 x 10^6	0.25 ppm
H_2O	1.25 x 10^3	108 x 10^6	0.1 ppm
NO_x (as NO_2)	10	860 x 10^3	2 ppb
Co	3	258 x 10^3	0.25 ppb
Hydrocarbons	0.5	43 x 10^3	43 ppt

Assuming that a residence time of 1 year applies to these exhaust products, the increase of mixing ratio at flight level (200 hPa) caused by these contaminants was calculated. The mass of the atmosphere was taken as 53 x 10^{14} t.

If the contaminants are spread evenly over the entire globe, the figures given in the last column of the table show the amount of disturbance to be expected.

The impact of CO_2 is just a contribution to the total amount of CO_2 released by burning fossil fuels. It has no specific impact on the chemistry of the atmosphere.

Vertical profiles of H_2O in the stratosphere are shown in fig. 5. If 0.1 ppm would be added at tropopause level, local saturation and thus cloud/contrail formation can be expected. The effect will be considerably enhanced if the emissions are not evenly distributed but rather concentrated in certain latitude regions. Increases of H_2O mixing ratios by 0.5 to 1 ppm may be expected there which make up 10 percent of the natural abundance of about 10 ppm at that altitude (see Fig. 5).

Contrail formations which are common in areas where flight corridors are concentrated, show that the water input from aircraft is significant.

Vertical profiles of NO and NO_2 are shown in figures 6 and 7. The increase of NO_x at the 200 hPa level due to aircraft emmission is 2 ppb. In regions of concentrated flight corridors it will certainly be higher. Compared to the natural background NO_x (Fig. 6 and 7) at 10 km this disturbance is enormous: NO_x abundances are enhanced by a factor 20. This large NO_x impact results in ozone increases by about 12 percent at this height (Hidalgo and Crutzen 1977). The steady increase of ozone observed at Hohenpeißenberg/Bavaria throughout the whole troposphere and lower stratosphere, up to about 15 km (Wege et al. 1989) may reflect, besides the impact of ground level emission, the additional ozone produced from NO_x emitted by aircraft.

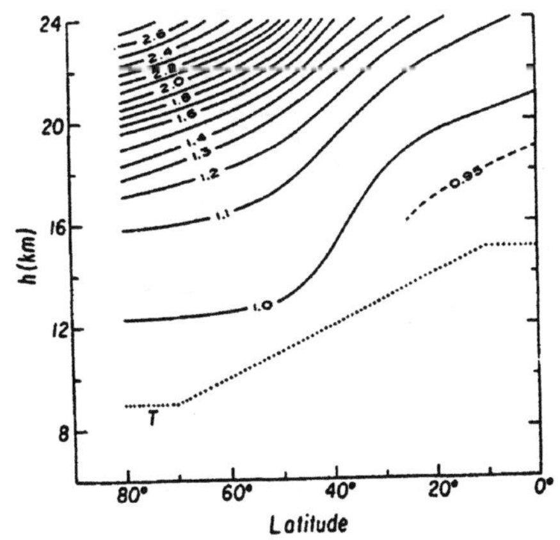

Figure 4: Stratospheric residence time (e-fold) as function of latitude and altitude. Units: years, T = tropopause (After Fabian 1974)

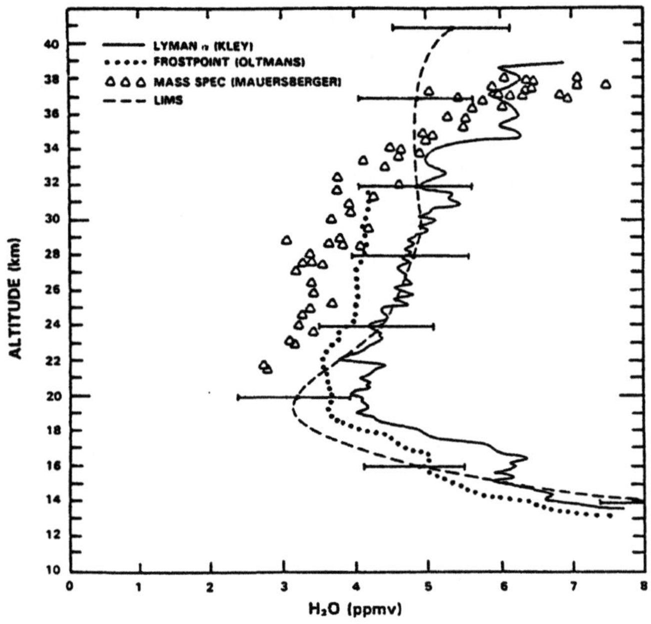

Figure 5: Vertical profiles of water vapour from various measurements (After WMO WMO 1986)

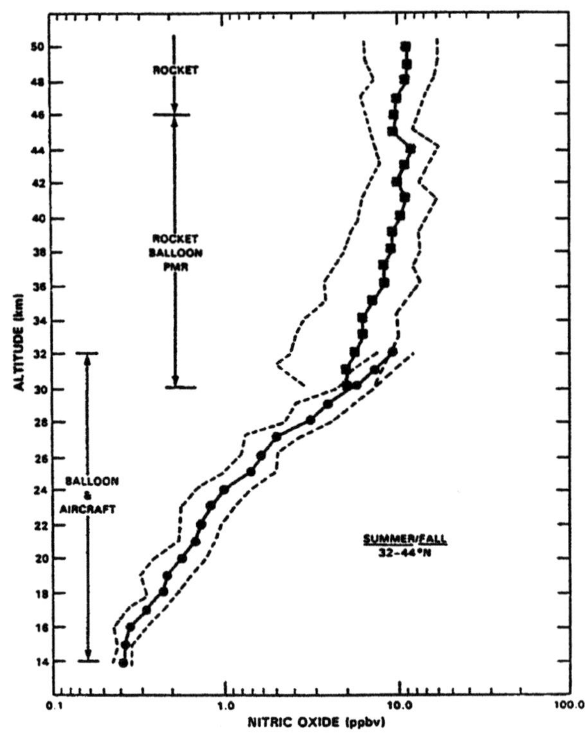

Figure 6: Vertical profiles of nitric oxide from various measurements (After WMO WMO 1986)

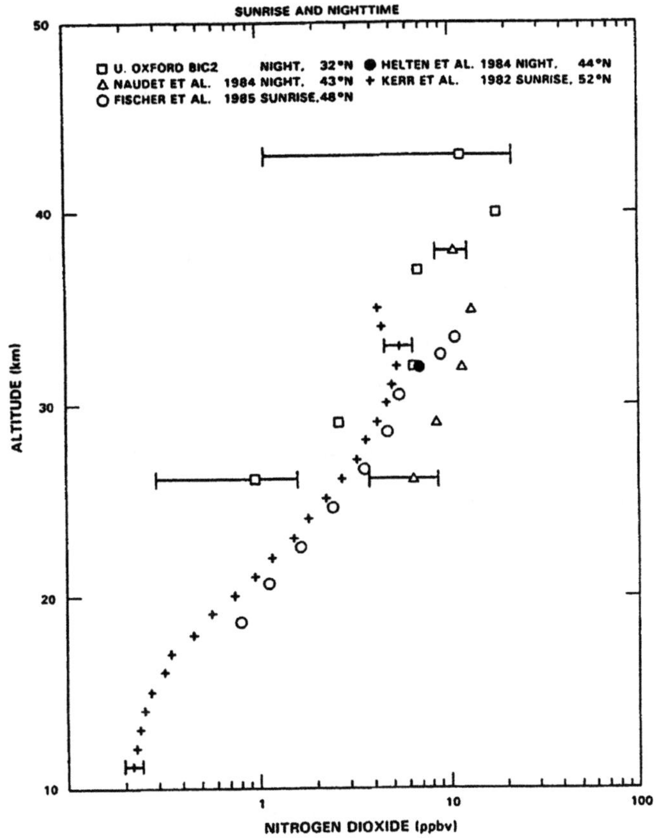

Figure 7: Vertical profiles of nitrogen dioxide from various measurements (After WMO 1986)

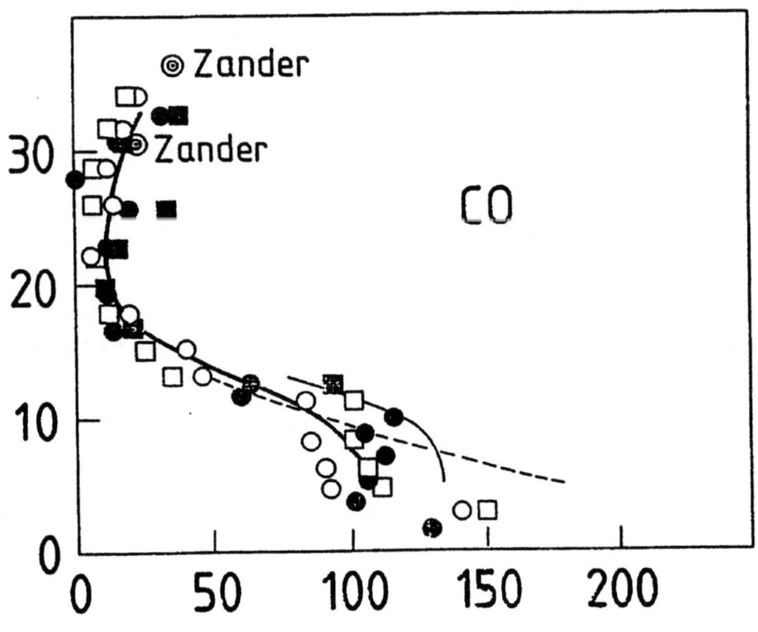

Figure 8: Vertical distribution of carbon monoxide mixingratio (ppb) as function of height (km) (After Fabian et al. 1981)

Carbon monoxide emissions cause increases of the CO mixing ratio of about 0.25 ppb, if all aircraft are evenly distributed over the globe. For regions of concentrated flight corridors this disturbance may be as high as 1 ppb. Compared to the background distribution of CO (fig. 8) this disturbance does not appear to be significant.

The effect of hydrocarbons is difficult to assess as only very few and sporadic background measurements exist. Rudolph and Ehhalt (1981) found considerable enhancements for mixing ratios of C_3H_8 and i-C_4H_{10} at the 200 hPa level which might be attributed to aviation. As hydrocarbons play a important role in photochemical smog reactions, more related measurements and model experiments are urgently needed.

This brief assessment shows that aircraft emissions, in particular H_2O and NO_x, may cause changes in atmospheric composition with possible consequences on ozone and climate. It is therefore important to study these effects in more detail.

REFERENCES

Chameides, W.L. and D.D. Davis: Chemistry in the troposphere. Chemical & Energineering News 60-40, 39, 1982

Fabian, P.: The effect of SST on the stratospheric distribution of odd nitrogen. Pure and Appl. Geophysics 112, 901-913, 1974

Fabian, P.: Verhalten der Flugzeug-Emissionen in der Luft: Umwandlungsprozesse. Ökologische Folgen des Flugverkehrs (M. Held, Ed.), Tutzinger Materialie 50/1988, 58-67, Evangelische Akademie Tutzing, FRG, 1988

Fabian, P.: Atmosphäre und Umwelt Springer-Verlag, 3. Auflage 1989

Fabian, P., R. Borchers, G. Flentje, W.A. Matthews, W. Seiler,

H. Giel, K. Bunse, F. Müller, U. Schmidt, A. Volz, A. Kedim, F.J. Johnen: The vertical distribution of stable trace gases at midlatitudes. J. Geophys. Res. 86, 5179, 1981

Hidalgo and P.J.J. Crutzen: J. Geophys.Res. 82, 4370, 1977

Oliver, Bauer Hidalgo, Gardner, Wasylkiwsky: HAPP Report 1977, US Dept. of Transportation, Washington, DC. 1977

Rudolph, J. and D.H. Ehhalt: Measurements of C_2-C_5 hydrocarbons, over the North Atlantic. J. Geophys. Res. 86, 11959, 1981

Telegadas, K.: The radioactivity distribution in the stratosphere from high yield Chinese and French unclear tests (1967-1970), Manuscript 1973

Wege, K., H. Claude and R. Harmannsgruber: Several results from 20 years of ozone observations at Hohenpeißenberg. Ozone in the Atmosphere (R.D. Bojkov, P. Fabian, Eds.) 109-112, Deepak Publishing, Hampton, Va. 1989

WMO: Atmospheric Ozone 1985. Report No. 16 I-III, NASA-FAA-NOAA-UNEP-WMO-CEC-BMFT, Washington, D.C. 1986

THE ATMOSPHERIC CHEMICAL EFFECTS OF AIRCRAFT OPERATIONS

Paul J. Crutzen
and
Christoph Brühl
Max–Planck–Institute for Chemistry
Airchemistry Department
P.O. Box 3060, D–6500 Mainz, F.R.G.

Abstract

The effect of aircraft emissions on atmospheric ozone has been studied with a two–dimensional model of the troposphere and stratosphere. Emissions in the first place of NO_x, but also of H_2O are the main factors influencing ozone. However, because of the growing presence of artificial chlorine compounds in the atmosphere important, but complex chemical interactions take place between all these compounds and their chemical derivatives, which must be considered for future projections of effects. In this paper we show that the effects are also strongly dependent on the altitude of emissions and discuss the implications of future H_2O and NO_x additions for polar stratospheric cloud formation and "ozone hole" reactions.

Introduction

Following the suggestion by Crutzen (1970) that NO and NO_2 could catalytically break down ozone, Johnston (1971) and Crutzen (1971, 1972) drew attention to the possibility that large fleets of supersonic aircraft could inject such massive amounts of NO into the stratosphere that substantial depletions of stratospheric ozone could occur. The large fleets of SST's were never built. However, a substantial expansion of subsonic aircraft operations in the upper troposphere and lower stratosphere has occurred during the past two decades, and is expected to continue or even accelerate in the future. Furthermore, plans are being developed to launch operations of advanced supersonic and hypersonic aircraft, some flying at appreciably higher altitudes than the 16–21 km planned originally for the SST's.

Although in the 1970's extensive studies were devoted to estimating the impact of aircraft operations on atmospheric ozone (e.g. CIAP, 1975; National Academy of Sciences, 1975; Hidalgo and Crutzen, 1977; Oliver et al, 1978) only recently there has been renewed interest in the issue. Scientific progress over the past twenty years has also shown that the issue of the chemical effects of stratospheric aircraft pollution is much more complex than originally thought, involving interactions between the NO_x, HO_x, and strongly growing amounts of ClX compounds, the latter produced by the chlorofluorocarbons produced by the chemical industry. In this paper we will give a brief overview of the current knowledge of the main chemical processes affecting atmospheric ozone and the impact of aircraft on it. We will also present some new results on the potential atmospheric chemical and climatic effects of future aircraft operations (the latter not in this text, but at the oral presentation).

Chemistry of atmospheric ozone

Stratospheric ozone is formed at altitudes above about 20 km by the photodissociation of molecular oxygen and recombination of the resulting oxygen atoms with an oxygen molecule:

$$3\,O_2 + h\nu \rightarrow 2\,O_3,\ \lambda < 240 nm \qquad (1)$$

More than 99% of the ozone molecules are in the stratosphere reconverted into oxygen molecules by a large variety of reaction chains whose net results can generally be written:

$$2\,O_3 + h\nu \rightarrow 3\,O_2 \qquad (2)$$

These reaction chains are catalyzed by the presence of various radical species in the stratosphere. Examples of such catalytic chains are:

$$\begin{aligned} OH + O_3 &\rightarrow HO_2 + O_2 \\ HO_2 + O &\rightarrow OH + O_2 \\ O_3 + h\nu &\rightarrow O + O_2 \\ \text{Net}: 2\,O_3 + h\nu &\rightarrow 3\,O_2 \end{aligned} \qquad (2a)$$

$$\begin{aligned} NO + O_3 &\rightarrow NO_2 + O_2 \\ NO_2 + O &\rightarrow NO + O_2 \\ O_3 + h\nu &\rightarrow O + O_2 \\ \text{Net}: 2\,O_3 + h\nu &\rightarrow 3\,O_2 \end{aligned} \qquad (2b)$$

$$\begin{aligned} Cl + O_3 &\rightarrow ClO + O_2 \\ ClO + O &\rightarrow Cl + O_2 \\ O_3 + h\nu &\rightarrow O + O_2 \\ \text{Net}: 2\,O_3 + h\nu &\rightarrow 3\,O_2 \end{aligned} \qquad (2c)$$

$$\begin{aligned} OH + O_3 &\rightarrow HO_2 + O_2 \\ HO_2 + O_3 &\rightarrow OH + 2\,O_2 \\ \text{Net}: 2\,O_3 + h\nu &\rightarrow 3\,O_2 \end{aligned} \qquad (2d)$$

$$\begin{aligned} Cl + O_3 &\rightarrow ClO + O_2 \\ Cl + O_3 &\rightarrow ClO + O_2 \\ ClO + ClO + M &\rightarrow Cl_2O_2 + M \\ Cl_2O_2 + h\nu &\rightarrow 2\,Cl + O_2 \\ \text{Net}: 2\,O_3 + h\nu &\rightarrow 3\,O_2 \end{aligned} \qquad (2e)$$

The catalytic cycle (2a), and some other analogous ones, dominates ozone removal above about 45 km; reaction cycle (2b) is most important between about 20 and 45 km, i.e. in the altitude region where most ozone is located; cycle (2c) is becoming of increasing importance in the same altitude region; cycle (2d) is important below 20 km; and (2e) is the main cycle that is responsible for the development of the ozone hole.

All of the listed catalytic chains are growing in importance due to the increasing abundance of the catalytic compounds. The reason for these increases is that the parent molecules H_2O, N_2O, CF_2Cl_2 and $CFCl_3$, which are responsible for the birth of the radical species are all increasing in the atmosphere, leading to enhanced radical species production via

$$\begin{aligned} O_3 + h\nu &\rightarrow O(^1D) + O_2 \quad \lambda < 310nm \\ O(^1D) + H_2O &\rightarrow 2\,OH \\ O(^1D) + N_2O &\rightarrow 2\,NO \end{aligned}$$

and

$$CF_iCl_j + h\nu + .. \rightarrow j\,Cl + CO_2 + i\,HF \quad \lambda < 270nm$$

The stratospheric abundance of H_2O is increasing, because it is produced by the oxidation of CH_4 ($CH_4 + 2O_2 \rightarrow CO_2 + 2H_2O$) whose atmospheric abundance is growing by about 1%/yr (WMO 1990). The growth in atmospheric N_2O is about 0.25%/yr (WMO 1990) and the CFC gases gases are increasing by about 4%/yr (in the average). Despite recent international agreements to halt the production of CFC gases, the active chlorine content of the stratosphere is expected to rise until about the year 2005 (from the model used in this study). Because the emissions of CH_4 and N_2O are intrinsically related to fossil fuel burning and food production, there is little doubt that their atmospheric abundances will continue to rise far into the next century. Any emissions of NO and H_2O from aircraft would thus enhance the buildup of these compounds which would anyhow take place.

The ozone depletion effects of the increases in the HO_x, NO_x, and ClO_x radical concentrations in the stratosphere are not additive. Chemical interactions between the radicals have to be considered. For instance, through the formation of $ClONO_2$ and HCl, which do not react with O_3, via the reactions

$$ClO + NO_2 + M \rightarrow ClONO_2 + M$$

and

$$ClO + NO \rightarrow Cl + NO_2$$
$$Cl + CH_4 \rightarrow HCl + CH_3$$

catalytic ozone depletion by NO_x and especially by ClO_x is being reduced. This also means that increasing concentrations of CH_4 tend to counteract ozone depletion by ClO_x catalysis. Partial ozone protection also occurs through the formation of HNO_3 and HNO_4 via

$$OH + NO_2(+M) \rightarrow HNO_3(+M)$$
$$HO_2 + NO_2 + M \rightarrow HNO_4 + M$$

On the other hand, reaction of HCl with OH, leading to Cl atoms

$$HCl + OH \rightarrow Cl + H_2O$$

enhances ozone depletions.

As another example of a feedback, ozone depletion by the catalytic set (2a) is short-circuited by a reaction of NO with HO_2:

$$OH + O_3 \rightarrow HO_2 + O_2$$
$$HO_2 + NO \rightarrow OH + NO_2$$
$$NO_2 + h\nu \rightarrow NO + O$$
$$O + O_2 + M \rightarrow O_3 + M$$
Net: no chemical effect (i.e. ozone protection),

thus counteracting the increased effect of (2b). This is especially important below 20 km. Furthermore in the troposphere, net ozone formation takes place in the oxidation cycles of CO and CH_4, e.g. via:

$$CO + OH \rightarrow CO_2 + H$$
$$H + O_2 + M \rightarrow HO_2 + M$$
$$HO_2 + NO \rightarrow OH + NO_2$$
$$NO_2 + h\nu \rightarrow NO + O$$
$$O + O_2 + M \rightarrow O_3 + M$$
$$\text{Net: } CO + 2O_2 \rightarrow CO_2 + O_3$$

and

$$CH_4 + OH \rightarrow CH_3 + H_2O$$
$$CH_3 + O_2 + M \rightarrow CH_3O_2 + M$$
$$CH_3O_2 + NO \rightarrow CH_3O + NO_2$$
$$CH_3O + O_2 \rightarrow CH_2O + HO_2$$
$$CH_2O + h\nu \rightarrow CO + H_2$$
$$HO_2 + NO \rightarrow OH + NO_2$$
$$NO_2 + h\nu \rightarrow NO + O \qquad (2\times)$$
$$O + O_2 + M \rightarrow O_3 + M \qquad (2\times)$$
$$\text{Net}: CH_4 + 4O_2 \rightarrow CO + H_2 + H_2O + 2O_3$$

All these, and other reactions not listed here, but considered in our study, have to be taken into account when making estimates of ozone depletions due to various anthropogenic activities. In general we can summarize the interactions as follows:

- H_2O and HO_x slow down NO_x, but enhance ClO_x catalysis.
- N_2O and NO_x slow down ClO_x and HO_x catalysis.
- NO_x enhances ozone depletion above about 20 km, but protects ozone from destruction and even produces it below 20 km.

The above reactions only consider gas phase reactions. The present development of the stratospheric "ozone hole" during springtime months over Antarctica has shown, that O_3 depletion by ClO_x catalysis is stimulated strongly by the removal of NO_x and HNO_3 due to the formation of NAT (nitric acid trihydrate) and ice particles at temperatures below about -80°C and -90°C, respectively. Due to the removal of NO_x and reactions on the particles, such as

$$ClONO_2 + HCl \rightarrow Cl_2 + HNO_3(solid)$$
$$Cl_2 + h\nu \rightarrow 2Cl$$

the inactive chlorine species HCl and $ClONO_2$ are converted into Cl atoms, establishing the powerful catalytic ozone depletion reaction chain (2e). With growing concentrations of NO_x due to increasing N_2O and NO_x from aircraft emissions, and of H_2O due to growing methane abundance and aircraft emissions, the likelihood of stratospheric particle formation, and ClO_x activation will increase. This is another factor to be considered in establishing the future impact of aircraft operations in the stratosphere.

Model results

A time dependent two dimensional (altitude/latitude) chemical dynamical model based on Gidel et al (1983), but with updated dynamics and chemistry, is used to estimate the effects of actual and future aircraft emissions in troposphere and stratosphere. Three scenarios were run from the year 1975 to 2025:

- case I without aircraft emissions,
- case II including aircraft emissions of NO_x and H_2O in the troposphere and lower stratosphere following the extrapolations of Wuebbles et al (1984), which do well in describing the current development, and
- case III, where in addition it is assumed, that after 2010 a fleet of 500 hypersonics may operate in a latitude belt between 30 and 50°N at an altitude of 27 km (Johnston et al, 1989).

In all scenarios it is assumed, that CH_4 and N_2O will continue to increase with the present growth rates (0.017ppm/yr and 0.25%/yr, respectively). The production of fully halogenated chlorofluorocarbons is assumed to be phased out after the year 2000 following the London amendment of the Montreal protocol, but including legal exemptions. CH_3CCl_3 will be phased out after 2005 while CFC-22 will be used temporarily as substitute until 2040. Industrial emissions of NO_x near may with its adverse effects to the biosphere.

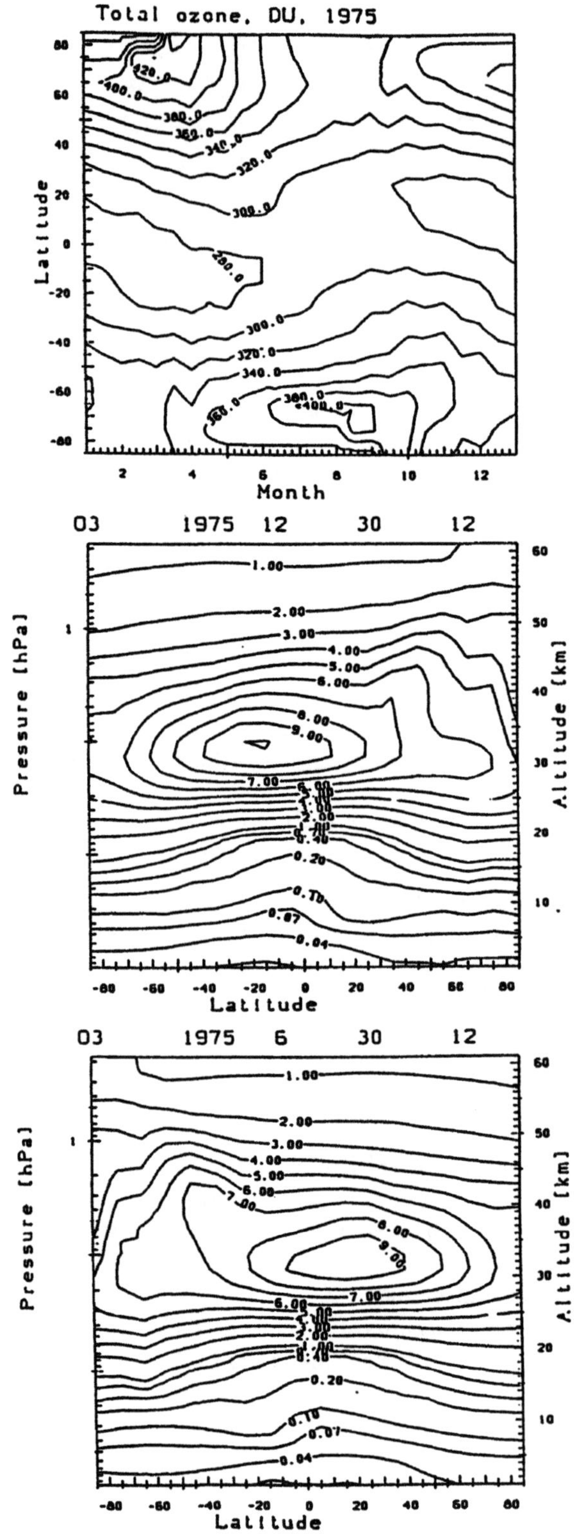

Figure 1: **Calculated total ozone in Dobson Units (panel a) and vertical distribution of ozone volume mixing ratio in *ppmv* in NH winter (panel b) and NH summer (panel c) of the year 1975.**

double from 1985 to 2030 following the assumptions of Kavanaugh (1987) for developing countries while surface CO emissions stay about constant. Aircraft emissions of CO are neglected.

Figure 1 shows the calculated seasonal and latitudinal distribution of total ozone and the vertical distribution of ozone in December and June for the year 1975 without perturbations by aircraft and only small perturbations by anthropogenic chlorine (ClX concentration about 1.4 ppbv compared to natural background of 0.6 ppbv). The development until now is given in Figure 2: the three panels on the lefthand side show the ozone perturbations in 1990 compared to 1975 without considering aircraft emissions, the panels on the righthand side include the effects of aircraft. The increase of active chlorine (to about 3 ppbv ClX) causes an ozone depletion by about 15% near 40 km in the winter hemispheres. The ozone increase in the troposphere due to increase in CH_4 and NO_x-emissions near the surface is enhanced by aircraft emissions especially in northern hemispheric summer. This effect causes increases in total ozone by up to almost 1% and counterbalances partially the depletion of stratospheric ozone by chlorine.

As can be seen from Figure 3, the differences in ozone with and without aircraft emissions of subsonics increase in the future, as shown for the year 2025. With aircraft emissions, ozone increases by more than 35% are calculated in the middle troposphere in northern summer, compared to 1990. Calculated total ozone increases everywhere compared to 1990 due to the increases in tropospheric ozone and because of the reductions in emissions of chlorine containing gases.

In case III, the depletion of stratospheric ozone by hypersonic aircraft with maximum values of 35% in winter at 28 km causes worldwide depletions in total ozone, which are most severe with up to 15% in northern spring (Figure 4). In the area, where the emissions occur, the mixing ratios of $NX = NO_x + HNO_3$ and H_2O increase by more than 150 and 100%, respectively (Figure 5, lower panels). Hypersonic aircraft cause an increase of NX by more than 50% in the whole lower stratosphere (below about 35 km or 10hPa) and increases of H_2O by more than 50% in the whole upper stratosphere (above about 30 km). The contributions to the NX and H_2O increases from the growths in N_2O and CH_4 concentrations are in this case relatively small compared to those coming from aircraft emissions (upper panels of Figure 5). Subsonic aircraft cause slight increases in stratospheric NX, but substantial NX increases by more than 100% in the tropopause region of the northern midlatitudes.

Conclusions

High-flying aircraft will cause substantial increases in lower stratospheric NX and H_2O concentrations. These increases may cause local ozone depletion in northern midlatitudes exceeding 30%, even if only gas-phase reactions are considered. The catalytically active substances and their precursors spread into the whole stratosphere, causing a significant depletion of total ozone worldwide, even if the emissions occur only in the mid-latitude belt in the northern hemisphere. Because ClX is supposed to decrease after about 2005, the leading role in future anthropogenic stratospheric ozone depletions may well be taken by enhanced NO_x emissions from aircraft.

Increases of the water vapor content by 50% and HNO_3 by 150% (Figure 5) in the altitude regions near 21 km, the regions, where in northern hemispheric winter polar stratospheric clouds (PSCs) were found, would significantly increase the likelihood for formation of these clouds, because for these conditions condensation would occur at higher temperatures (Hanson and Mauersberger, 1988). Since in polar stratospheric clouds the chlorine reservoir species HCl and $ClONO_2$ are converted to its ozone destroying forms, substantial additional ozone depletions in the lower stratosphere of the northern hemisphere may occur. Subsonic aircraft have also some effects on the likelihood of PSC-formation, however, they are small because of the calculated HNO_3 increase of only about 10% in the critical region near 21 km.

Subsonic air traffic may cause a substantial increase in upper tropospheric and lower stratospheric ozone, which is an important greenhouse gas (e.g. Fishman et al., 1979; Brühl and Crutzen, 1988). Tropospheric ozone production counteracts the stratospheric ozone depletions, however, the calculated increases of about 10 to 15% near the surface contribute to the photochemical smog

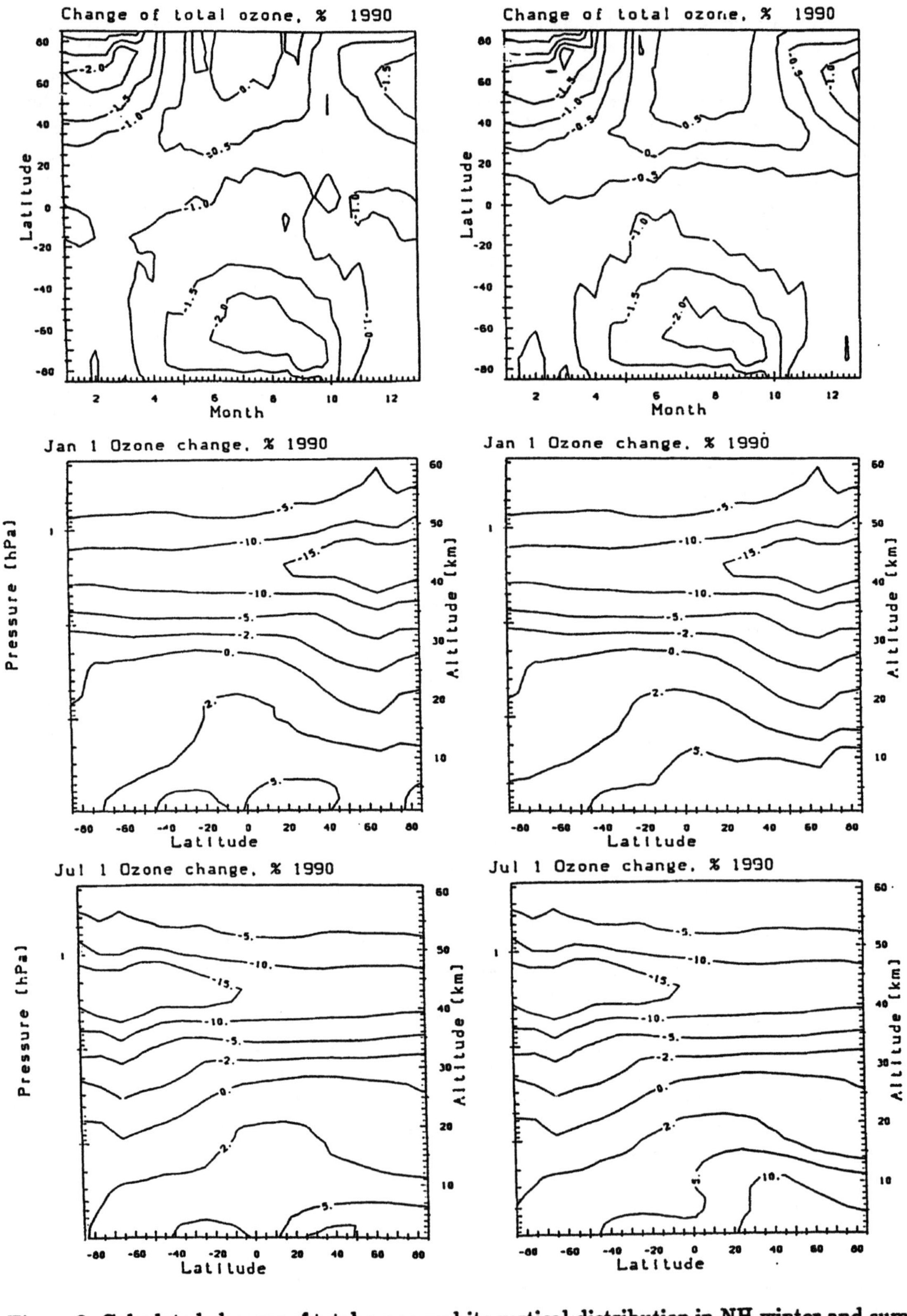

Figure 2: **Calculated changes of total ozone and its vertical distribution in NH winter and summer in 1990 compared to 1975. Lefthand side: without aircraft emissions; righthand side with aircraft emissions.**

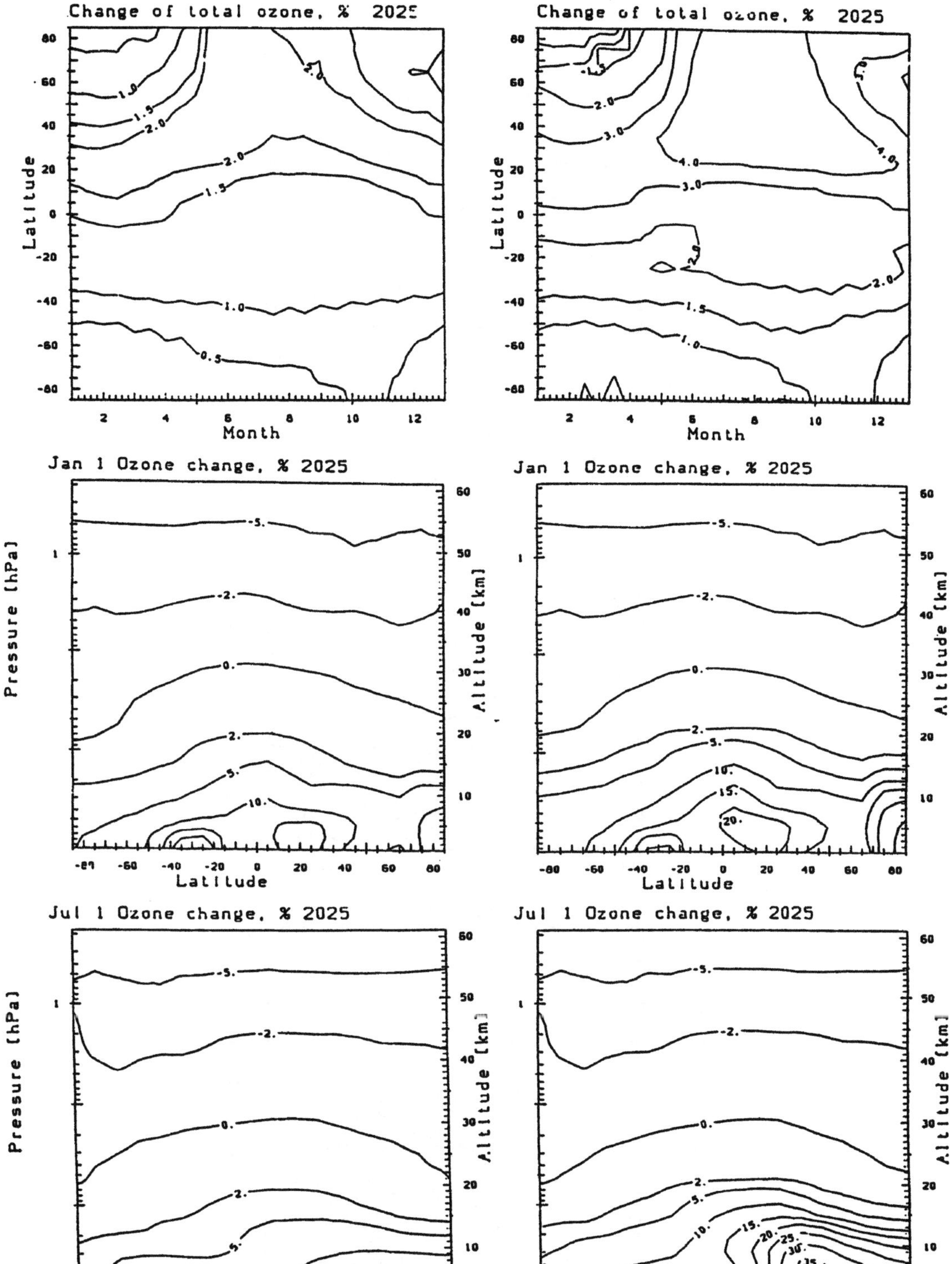

Figure 3: **Calculated changes of total ozone and its vertical distribution in NH winter and summer in 2025 compared to 1990 without (lefthand side, case I) and with (righthand side, case II) aircraft emissions.**

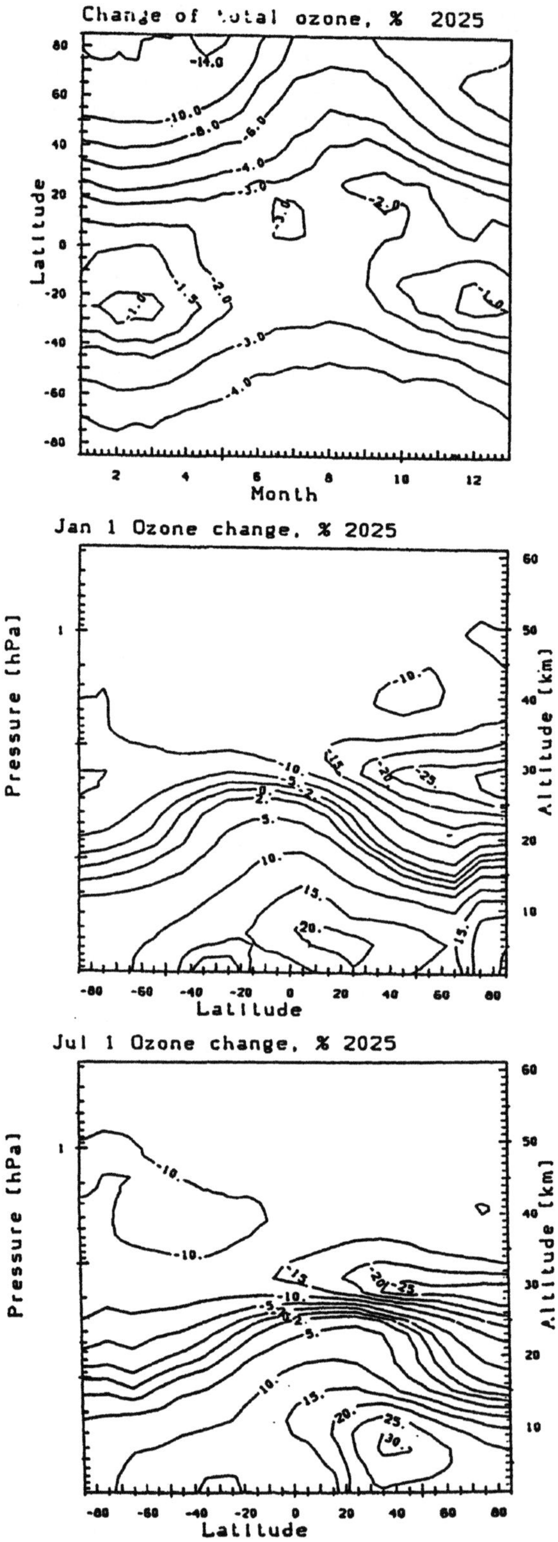

Figure 4: **As Figure 3, but for case III (with sub- and hypersonics).**

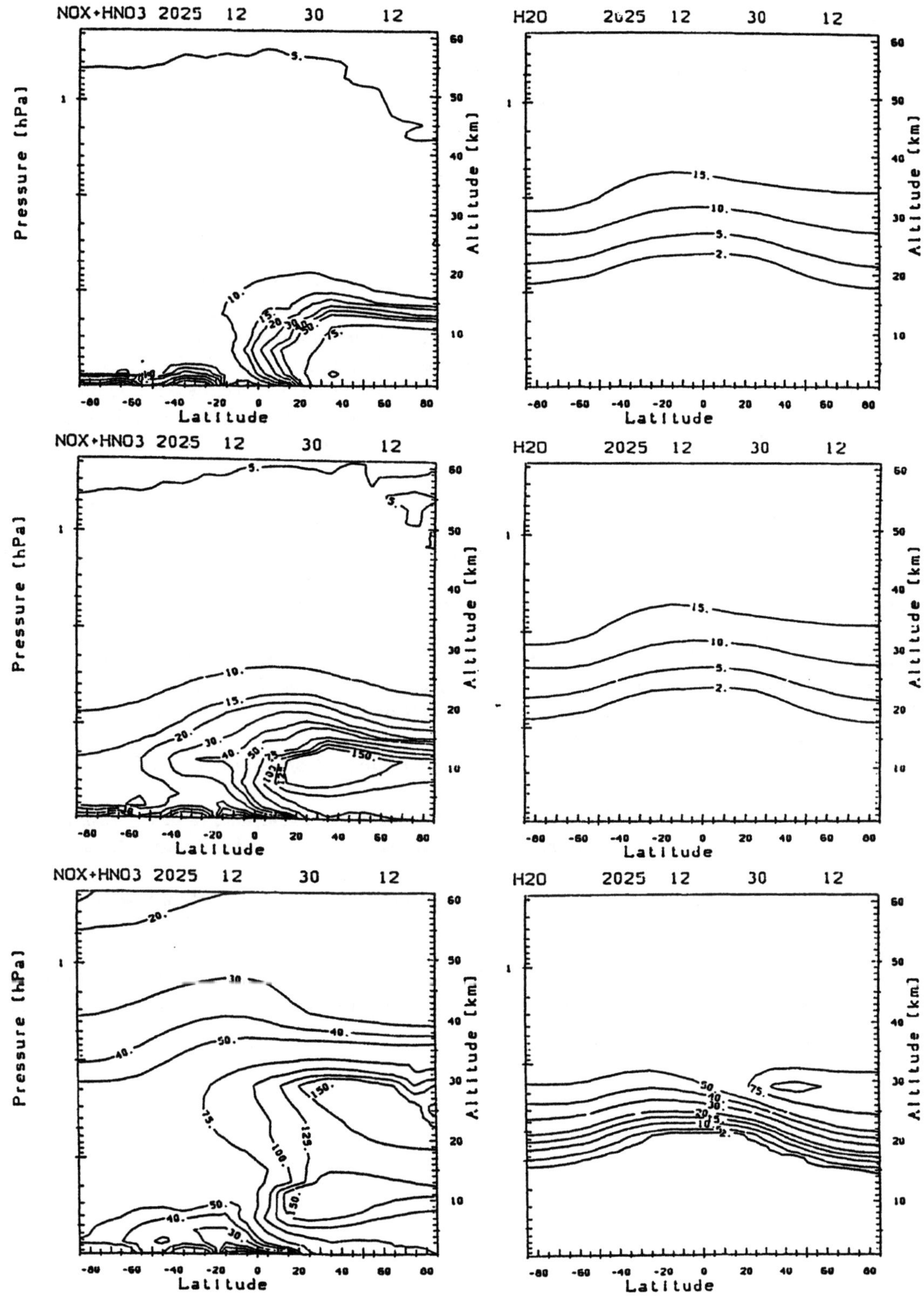

Figure 5: **Increases in $NX = NO_x + HNO_3$ and H_2O concentrations in 2025 (northern winter) compared to 1990 for the three scenarios: upper panels: I, no aircraft; middle: II, subsonics only; lower panels: III: subsonis and hypersonics.**

<u>*References*</u>

Brühl, C., P.J. Crutzen, 1988: Scenarios of possible changes in atmospheric temperatures and ozone concentrations due to man's activities, estimated with a one-dimensional coupled photochemical climate model. *Climate Dynamics* **2**, 173-203.

CIAP, 1975: *The stratosphere perturbed by propulsion effluents.* CIAP Monograph III, Report TST-75-53, US-Department of Transportation, Washington, DC.

Crutzen, P.J., 1970: The influence of nitrogen oxides on the atmospheric ozone content. *Quart.J.Roy.Met.Soc.* **96**, 320-325.

Crutzen, P.J., 1971: Ozone production rates in an oxygen-hydrogen-nitrogen oxide atmosphere. *J.Geophys.Res.* **76**, 7311-7327.

Crutzen, P.J., 1972: The photochemistry of the stratosphere with special attention given to the effects of NO_x emitted by supersonic aircraft. In: CIAP Proceedings of the survey conference (Ed. A.E. Barrington), Report TSC-OST-72-13, US-Department of Transportation, 80-89.

Fishman, J., S. Solomon, P.J. Crutzen, 1979: Observational and theoretical evidence in support of a significant in-situ photochemical source of tropospheric ozone. *Tellus* **31**, 432-446.

Gidel, L.T., P.J. Crutzen and J. Fishman, 1983: A two-dimensional photochemical model of the atmosphere. I: Chlorocarbon emissions and their effect on stratospheric ozone. *J.Geophys.Res.* **88**, 6622-6640.

Hanson, D., K. Mauersberger, 1988: Laboratory studies of the nitric acid trihydrate: implications for the south polar stratosphere. *Geophys.Res.Let.* **15**, 855-858.

Hidalgo, H. and P.J. Crutzen, 1977: The tropospheric and stratospheric composition perturbed by the NO_x emissions of high-altitude aircraft. *J.Geophys.Res.* **82**, 5833-5866.

Johnston, H.S., 1971: Reduction of stratospheric ozone by nitrogen catalysts from supersonic transport exhaust. *Science* **173**, 517-522.

Johnston, H.S., D.E. Kinnison and D.J. Wuebbles, 1989: Nitrogen oxides from high-altitude aircraft: an update of potential effects on ozone. *J.Geophys.Res.* **94**, 16351-16363.

Kavanaugh, M., 1987: Estimates of future CO, N_2O and NO_x emissions from energy combustion. *Atmos.Environment* **21**, 463-468.

National Academy of Sciences, 1975: *Environmental impact of stratospheric flight.* Washington, DC.

Oliver, R.C., E. Bauer and W. Wasylkiwskyj, 1978: *Recent developments in the estimation of the potential effects of high altitude aircraft emissions on ozone and climate.* Report No. FAA-AEE-78-24, US-Department of Transportation, Washington, DC.

WMO, 1990: *Scientific assessment of stratospheric ozone 1989.* World Meteorological Organization global ozone research and monitoring project. Report no 20. NASA, Washington, DC / WMO, Geneva.

Wuebbles, D.J., M.C. McCracken and F.M. Luther, 1984: *A proposed reference set of scenarios for radiatively active atmospheric constituents.* Report TR015 (W-7405-ENG-48), US-Department of Energy, Washington, DC.

SENSITIVITY OF STRATOSPHERIC OZONE TO PRESENT AND POSSIBLE FUTURE AIRCRAFT EMISSIONS

Donald J. Wuebbles and Douglas E. Kinnison
Atmospheric and Geophysical Sciences Division
Lawrence Livermore National Laboratory
Livermore, CA 94550

ABSTRACT

The aircraft industry is showing renewed interest in the development of supersonic, high flying aircraft for intercontinental passenger flights. There appears to be confidence that such high-speed civil transports can be designed, and that these aircraft will be economically viable as long as they are also environmentally acceptable. As such, it is important to establish the potential for such environmental problems early in the aircraft design. Initial studies with LLNL models of global atmospheric chemical, radiative, and transport processes have indicated that substantial decreases in stratospheric ozone concentrations could result from emissions of NO_x from aircraft flying in the stratosphere, depending on the fleet size and magnitude of the engine emissions. The purpose of this study is to build on previous analyses of potential aircraft emission effects on ozone in order to better define the sensitivity of ozone to such emissions. In addition to NO_x, the effects of potential emissions of carbon monoxide and water vapor are also examined. More realistic scenarios for the emissions as a function of altitude, latitude, and season are examined in comparison to prior analyses. These studies indicate that the effects on ozone are sensitive to the altitude and latitude, as well as the magnitude, of the emissions.

INTRODUCTION

The first research studies of the potential environmental impacts from commercial fleets of aircraft flying in the stratosphere were done in the early 1970s (Harrison,1970; Johnston, 1971). These studies suggested that the emissions from fleets of such aircraft could cause a significant reduction in the concentrations of ozone in the stratosphere, with accompanying increases in the amount of ultraviolet radiation reaching the Earth's surface. The primary concern, as pointed out by Johnston, was the chemical destruction of ozone from emissions of nitrogen oxides (NO_x) produced thermally in the aircraft engine combustion process. As a consequence of these studies, various governments instigated research programs to clarify the potential environmental effects of aircraft operations in the stratosphere. In the U.S., concurrent, but independent, studies were conducted by the Department of Transportation's Climatic Impact Assessment Program (CIAP, 1974, 1975 a, b) and by the National Academy of Sciences (NAS, 1975). Several independent assessments were also carried out in Europe (COMESA, 1976; COVOS, 1976). A subsequent study to CIAP, the High Altitude Pollution Program (HAPP) sponsored by the U.S. Department of Transportation, continued studies on potential aircraft effects at a lower research level until 1980.

The basic conclusions of these assessment studies were similar, but with some differences in the estimated magnitudes of the environmental effects. Large-scale aircraft operations in the stratosphere were concluded to lead to significant and potentially unacceptable levels of ozone destruction. Most of the ozone reduction was estimated to be caused by emissions of nitrogen oxides with a small effect due to water vapor emissions. However, the small fleet (30 or less) estimated for Concorde-type supersonic transport was determined to have little impact on global ozone or climate.

The primary research tool for investigating the potential effects of aircraft emissions on stratospheric throughout the periods of CIAP and HAPP was the one-dimensional (1-D) chemical-radiative-transport model of the global atmosphere. Such models, including the one at LLNL, determine the vertical distributions of the important trace constituents in the atmosphere, and have the advantage of being computationally efficient while including detailed representations of atmospheric chemical and radiative processes. However, the treatment of atmospheric dynamical processes in these models through an empirically based eddy diffusion representation has well-recognized limitations (WMO, 1985).

Although the two-dimensional (latitude and altitude) chemical-radiative-transport model to be described below is now our primary research tool for stratospheric chemistry studies, we have continued to maintain a baseline of the calculated one-dimensional model effects on ozone for the standard CIAP aircraft scenario (as well as for a standard chlorofluorocarbons, CFCs, emissions scenario, as shown in Figure 1). The results in Figure 1 show the effects of the maturing of stratospheric science from the time of the early CIAP analyses when stratospheric chemistry was comparatively poorly understood. The effects on ozone for this scenario varied considerably throughout the 1970s, even going positive for a two year period while the understanding of hydrogen (HO_x) chemistry was being extensively scrutinized in laboratory kinetics studies. By 1988, the calculated ozone reduction by NO_x emissions at 20 km (the peak altitude of the original proposed U.S. SST) was about the same as it was at the conclusion of CIAP in 1974. The current model gives essentially the same result as the 1988 version.

A number of countries have shown renewed interest in the development of fast, high flying aircraft for intercontinental passenger flights. Such high speed civil transports (HSCTs) could reduce international travel time by a factor of two or more. The aircraft industry appears to be confident that such high-speed civil transports can be designed, and that these aircraft will be economically viable as long as they are also environmentally acceptable. As such, it is important to establish the potential for such environmental problems early in the aircraft design.

The first detailed sensitivity analysis in almost a decade of the effects from aircraft emissions on ozone was published by Johnston et al., in 1989 (part of this work also appeared in Kinnison et al., 1988). Both the LLNL one-dimensional and two-dimensional (2-D) models of the global atmosphere were used in this study. The one-dimensional model provides a historical perspective along with having computational advantages for multiple sensitivity calculations, while the two-dimensional model provides information on seasonal and latitude effects as well as having a stronger theoretical basis for its representation of atmospheric transport processes. Major findings from this study were:

1. Nitrogen oxides from the exhaust gases of a fleet of HSCTs can reduce stratospheric ozone on a global basis. These calculated ozone reductions depend strongly on the altitude and magnitude of the NO_x injection. Figure 2 (based on the results of Johnston, et al., but not

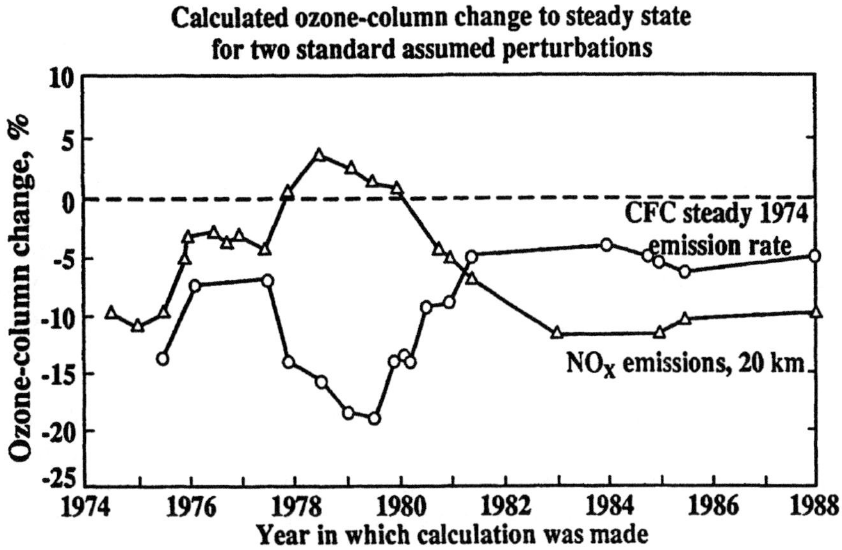

Figure 1. Calculated change in the total ozone column at steady state for two standard assumed perturbations: (a) the standard CIAP scenario for emissions of NO_x at 2000 molecules $cm^{-3}s^{-1}$ over a 1 km altitude interval centered at 20 km, and (b) continuous emissions of CFC-11 and CFC-12 at 1974 rates. The LLNL one-dimensional model was used in all calculations over this 14-year period; the calculations are based on the version of the model with chemistry and physics treatments that were either recommended or considered current for that particular year.

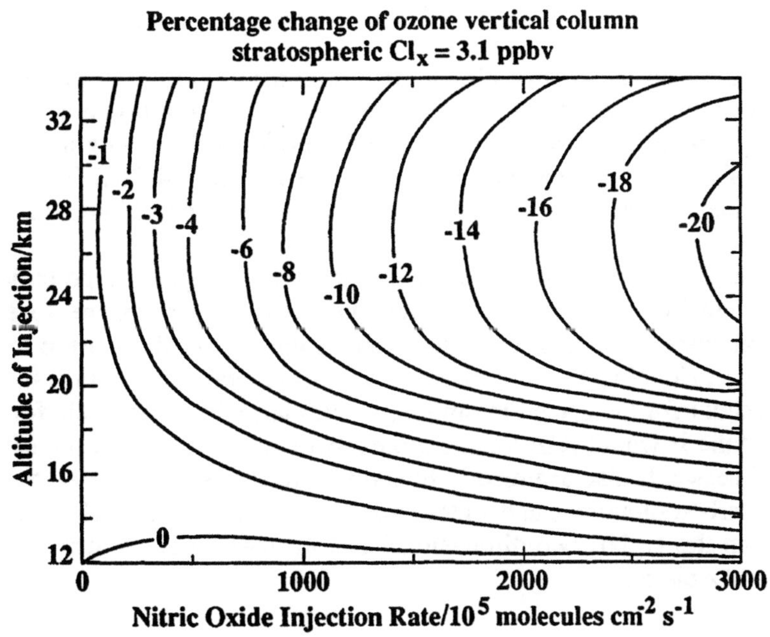

Figure 2. Percentage change in the total ozone column as calculated with the LLNL one-dimensional model as a function of the altitude and magnitude of the NO_x injection. The emissions are in units of molecules $cm^{-3}s^{-1}$ over a 1 km band centered at the stated altitude. The percentage change in ozone is relative to a reference atmosphere that contains 3.1 ppbv of stratospheric chlorine (Cl_x).

presented in the paper) shows one-dimensional model results indicating that NO_x injections below about 13 km cause an increase in the total ozone column, while injections above 13 km result in a net destruction of ozone.

2. The altitude at which NO_x emissions cause the largest reduction in the ozone column is about 25 km in the one-dimensional model and about 28 km in the two-dimensional model.

3. For a given altitude and magnitude of NO_x emissions, the two-dimensional model indicates that the reduction of global ozone depends on the latitude of the injections, with the maximum ozone reduction for tropical injection. For a given injection, the largest ozone column reductions occur in the polar regions.

4. For extremely large stratospheric chlorine (Cl_x) mixing ratios (~8 ppbv compared to about 3 ppbv in the current atmosphere), NO_x emissions can increase the ozone column, partially counteracting the ozone reduction caused by the chlorine. This effect is only found in a highly Cl_x perturbed stratosphere where ozone is already greatly depleted.

5. Water vapor (without NO_x) emissions cause a small ozone reduction; water vapor and NO_x together give an ozone reduction less than that calculated for NO_x emissions alone by a factor between 0.85 and 0.97, depending on the altitude and magnitude of the emissions.

6. Calculated global reductions of ozone due to NO_x emissions are smaller in the two-dimensional model than in the one-dimensional model, by factors that range from 0.66 to 0.86 for injection altitudes between 20 to 34 km. [More recent calculations show a sensitivity in the two-dimensional model to uncertainties in the treatment of eddy transport processes (Kinnison, 1989).]

7. On the basis of an uncertain estimate of fuel consumption by future stratospheric aircraft, two-dimensional model calculations for a fleet of 500 aircraft operating at 22 km with NO_x emission properties of 1988 subsonic commercial aircraft engines give a global ozone reduction of 19% (assuming emissions occur between 37°–49°N). The effect of reducing the emission index for the amount of NO_x produced per kg of fuel burned is shown in Table 1.

The purpose of this study is to build on the previous analyses of potential aircraft emission effects on ozone in order to better define the sensitivity of ozone to such emissions. A preliminary analysis is made of the sensitivity of the ozone distribution to the current fleet of commercial aircraft. The future emissions scenarios evaluated in this study attempt to more accurately account for the spread in emissions with latitude and altitude for an assumed fleet of HSCTs. The time it would take for the atmosphere to recover from the effects on ozone after emissions are ceased is also examined. In addition to emissions of NO_x, the effects of potential emissions of carbon monoxide and water vapor are also evaluated. These analyses are all based on results from the LLNL two-dimensional chemical-radiative-transport model of the troposphere and stratosphere.

Table 1. Calculated change in global-averaged total ozone from the LLNL two-dimensional model assuming an annual fuel consumption of 7.7 x 10^{10} kg, with all emissions between 37°–49°N in a 3 km altitude range centered about 22.5 km altitude (relative to an atmosphere with no aircraft emissions), based on Johnston, et al. (1989).

Emission Scenario	Emission Index g of NO Kg^{-1} Fuel	Injection Rate MT of NO_2 Yr^{-1}	Injection Rate Molecules/ 10^5 $cm^{-2}s^{-1}$	Change in Total Ozone
Current Subsonic Rate	40*	4.8	4000	−19.0%
Standard CIAP Rate	15	1.8	1500	− 8.6%
Future Goal	5	0.6	500	− 2.8%

* This emission index assumes current commercial subsonic aircraft technology being used in the stratosphere. The actual emission index for current aircraft at their flight altitudes in the upper troposphere are a factor of 2–3 smaller than this.

THE LLNL TWO-DIMENSIONAL MODEL

The two-dimensional model of the global atmosphere provides a zonally-averaged representation of the chemical and physical processes determining the composition and distributions of trace constituents in the troposphere and stratosphere. The concentrations of 35 chemically active trace constituents, based on more than 100 chemical and photochemical reactions, are calculated as a function of altitude, latitude, and season within the model. Essentially the model determines the atmospheric distribution of ozone and other important constituents based on the interactions of chemical, radiative, and dynamical transport processes thought to be operating in the atmosphere. For the version of the model used here, the diabatically-driven circulation is determined for the current (background) atmosphere using observed temperatures along with model-derived radiative transfer calculations of the net solar and longwave heating, determined in an internally consistent way with the derived species distributions. For the perturbed atmosphere, a perturbation form of the thermodynamic equation is solved for the changes in stratospheric temperatures resulting from the changes in ozone and other radiatively important trace constituents (note that the diabatic circulation is assumed to be unchanged from the current atmosphere for these calculations). More detailed descriptions of the two-dimensional model can be found in Wuebbles and Kinnison (1989), Johnston, et al. (1989), and Kinnison (1989).

CURRENT AIRCRAFT EMISSIONS

Earlier calculations with the one-dimensional model (Wuebbles, 1983) suggested that emissions from the existing, primarily subsonic, fleet of commercial aircraft could affect upper tropospheric ozone, resulting in increased concentrations of ozone in the region just below the tropopause. However, the one-dimensional model is of limited use in evaluating changes in tropospheric chemistry. Even the two-dimensional model is somewhat limited in treating tropospheric chemistry, particularly in the lower troposphere, because of the significant impact of emissions from highly reactive species primarily produced over land; only a three-dimensional model can be used to fully evaluate the effects on the global atmosphere from such emissions. Nonetheless, the two-dimensional model provides a useful test of how sensitive the troposphere may be to aircraft emissions primarily occuring in the upper troposphere.

One of the difficulties in evaluating the effects of existing aircraft emissions is that there isn't an available evaluation of the trends in such emissions over the last several decades. The most complete estimate of current emissions comes from the analysis of commercial aircraft flights for 1987 in a recent analysis by Boeing Corporation (Boeing, 1989). Slightly more than 29,000 city-pairs were accounted for in their analysis, along with the appropriate aircraft types and flight frequencies. The analysis determined emissions of nitrogen oxides, carbon monoxide, and hydrocarbons at 26,000 feet (about 8 km) and 37,000 feet (about 11 km). Figure 3a shows a contour plot of the NO_x emissions from the Boeing analysis. Bauer (1978) published an estimate of the 1975 subsonic aircraft emissions that had about a factor of 3.4 less NO_x emitted than the assumed emissions for 1987. If this earlier emissions estimate is correct, there was a sizable increase in aircraft emissions during the 12 years between 1975 and 1987.

The two-dimensional model is used to evaluate the effects on ozone concentrations resulting from the 1987 fleet emissions compared to an ambient atmosphere with no aircraft emissions. Emissions of nitrogen oxides, carbon monoxide, and hydrocarbons were included in the calculation based on the Boeing analysis; the hydrocarbon emissions were assumed to have the same reactivity as methane, as there was no attempt in this preliminary study to represent the complexity of non-methane hydrocarbon chemistry in this version of the model.

The calculated response of the ozone distribution to the emissions for the 1987 fleet are shown in Figures 3b and 3c. The largest increase in ozone occurs near 9 km at midlatitudes in the Northern Hemisphere, corresponding to the altitude and latitude of the maximum emissions. Approximately a 5% maximum increase in ozone is determined in July, primarily due to the effect of the NO_x (see Table 2) emissions to produce ozone in the troposphere through smog-formation reactions (e.g., see WMO, 1985). The global-averaged change in total ozone is 0.45%, with the maximum change of 0.6% occurring at high latitudes in the Northern Hemisphere.

Table 2. Sensitivity analyses to determine the relative importance of NO_x, CO, and hydrocarbon (represented as CH_4 emissions) on the calculated change in total ozone for the current aircraft emissions scenario.

	Change in total ozone (%)		
	Global Average	Northern Hemisphere	Southern Hemisphere
$NO_x + CO + CH_4$	0.45	0.53	0.37
NO_x only	0.44	0.52	0.36
CO only	0.018	-0.0058	0.043
CH_4 only	0.024	0.0031	0.045

A trend analysis of ozonesonde measurements from nine stations at latitudes from 32°N to 75°N has been recently made for the years from 1965 through 1986 (WMO, 1989). Significant increases in ozone concentrations were determined throughout the troposphere, with the largest increase (+ 0.83% per year for the period from 1970 to 1986) occurring in the bottom 3 km of the atmosphere. A secondary peak in the ozone increase (+ 0.38% per year or 6% over the 16 year period of the trend estimate) was found at altitudes between 8 and 10 km, just below the tropopause. Although it is premature to attribute causality, it is interesting that the secondary peak occurs at the same altitude and magnitude as the calculated effect from subsonic aircraft emissions. Aircraft emissions may explain a substantial fraction of the observed ozone increase in the upper troposphere. However, one should be extremely cautious in interpreting the global implications of such a limited data set.

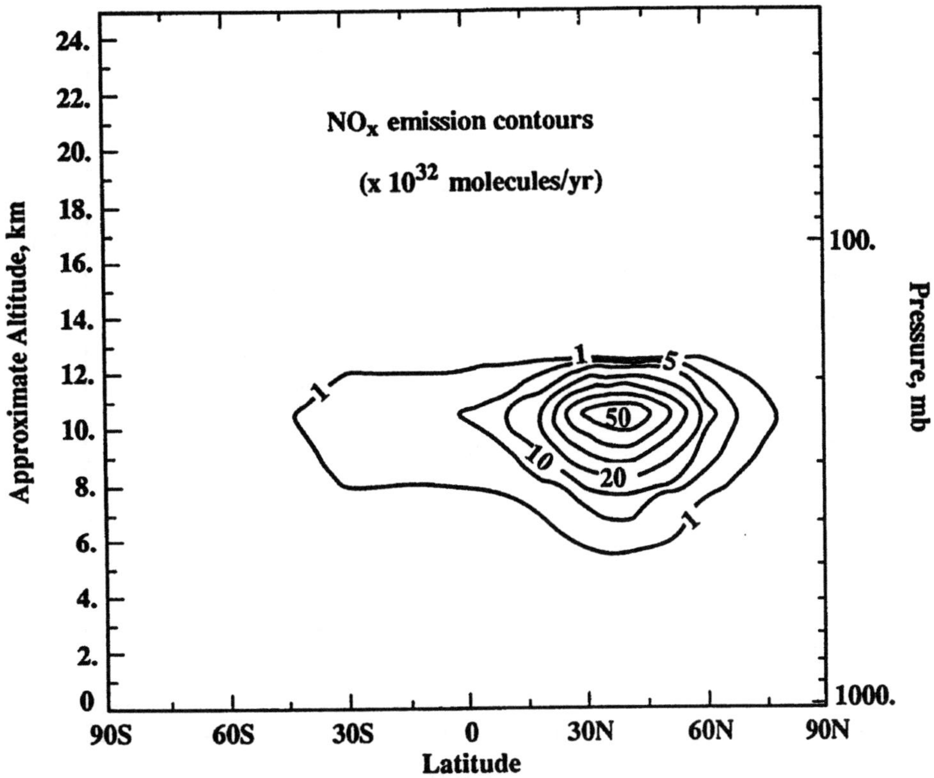

Figure 3(a)

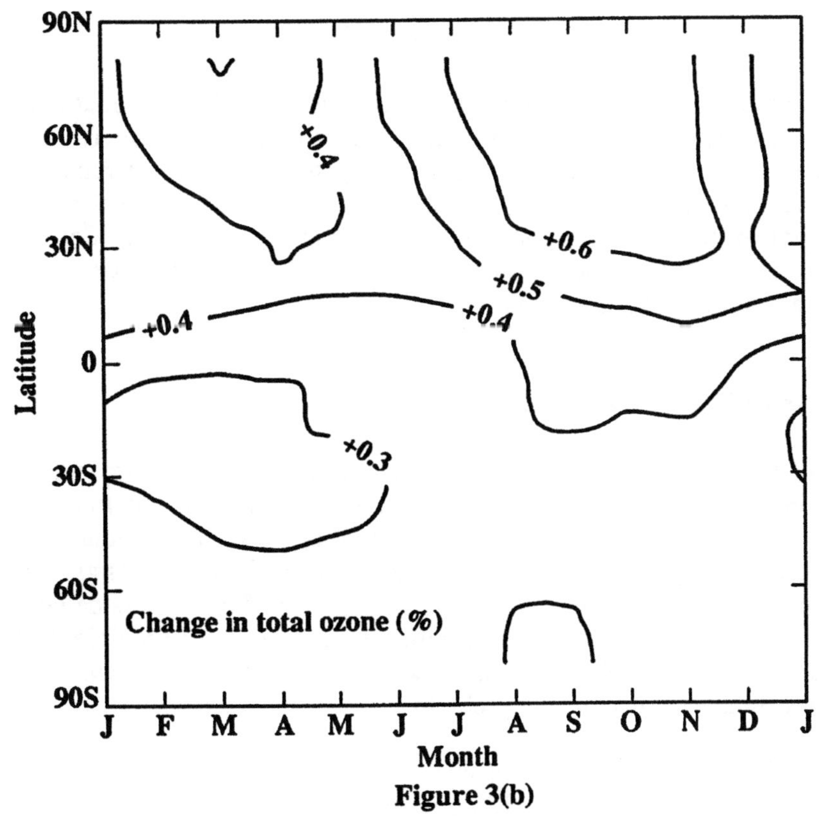

Figure 3(b)

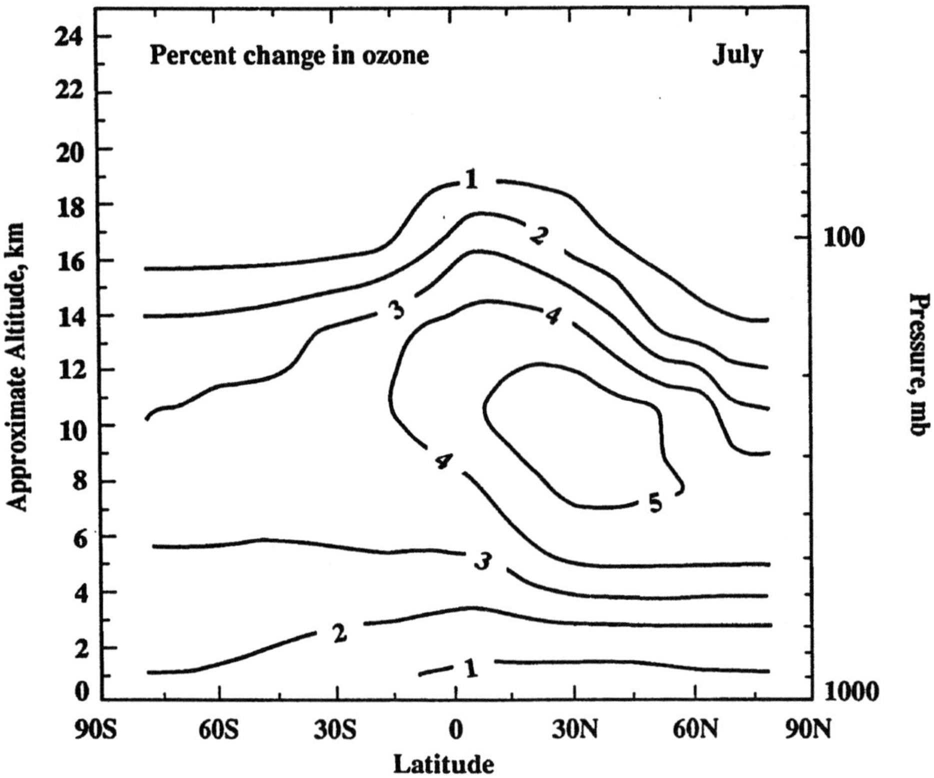

Figure 3(c)

Figure 3. (a) Contour of the NO_x emissions as a function of altitude and latitude determined by Boeing (1989) in their evaluation of the commercial aircraft emissions for the year 1987; (b) Calculated percentage change in total ozone at steady-state plotted as a function of latitude and time of the year as determined in the LLNL two-dimensional model for the 1987 subsonic aircraft fleet compared to an atmosphere without aircraft emissions; (c) Calculated percentage change in ozone at steady-state plotted as a function of altitude and latitude as determined for the 1987 subsonic aircraft fleet during the month of July .

A MATRIX OF FUTURE EMISSIONS

The amount of nitrogen oxides and other gases emitted into the atmosphere by a fleet of commercial aircraft depend on a number of factors, including the fleet size, engine fuel efficiency, the aircraft range, the cruise altitude, flight operational procedures, and the emission index for amount of NO_x produced per kg of fuel burned. Possible designs for HSCT aircraft and their engines are still at an early stage; therefore, most of the factors determining emissions of an actual fleet are still unknown. Examining a wide range of possible flight altitudes and levels of emissions for the effects on ozone will help establish the design criteria needed to prevent an unacceptable environmental problem.

In this study, we have evaluated a total of twelve HSCT fleet scenarios assuming three different cruise altitudes and varying levels of NO_x emissions. These hypothetical scenarios were developed by the HSCT group at McDonnell Douglas Corporation in a cooperative effort to establish a matrix of scenarios with realistic assumptions for flight altitudes versus magnitude of emissions. In no manner are these scenarios based on actual aircraft designs; they are intended to cover a broad range of aircraft altitudes and emissions These scenarios attempt to consider how the emissions should vary with latitude and altitude appropriate to a worldwide fleet of HSCTs. No subsonic aircraft emissions were considered in these scenarios. Three different aircraft, flying at different mean cruise altitudes, were assumed in the scenarios; the mean cruise altitudes were 15.8, 18.3, 22.9 km, although changes in flight altitudes throughout the flight patterns assumed in the scenarios are accounted for. Ten different city-pairs were chosen by McDonnell Douglas to represent the appropriate regions for intercontinental flights; these flight patterns combined with the assumed aircraft type provide the criteria for the altitude and latitude of the aircraft emissions. A wide range of NO_x emissions were assumed, based on uncertainties in fleet size, fuel burned, and NO_x emission index. All scenarios assume that a kerosene based fuel will be used in the HSCT in contrast to a more exotic fuel such as liquid methane or hydrogen.

Table 3 indicates the mean cruise altitude and the total global NO_x emissions for the 12 scenarios. Figure 4 shows the NO_x emissions assumed in scenario c1, corresponding to a mean flight altitude of 22.3 km, a total NO_x emission of 0.28 Mt per year (as NO_2) or 3.64×10^{33} molecules per year. The NO_x emissions for scenarios with the other mean cruise altitudes look very similar to Figure 4, except maximum emissions occur at lower altitudes.

For each of the 12 scenarios, Table 3 shows the calculated change in total ozone column for each hemisphere, along with the global averaged change in ozone. These calculations are based on a background stratospheric chlorine level of 2.9 ppbv, corresponding to current concentrations.

Table 3. Annually-averaged change in ozone for the twelve emissions scenarios evaluated with the LLNL two-dimensional model.

Scenarios	Mean Cruise Altitude (km)	Total NO_x Emissions ($\times 10^{33}$ Molecules/Yr)	ΔO_3/% N.H.	ΔO_3/% S.H.	ΔO_3/% Global
a0	15.8	0.67	− 0.20	−0.18	− 0.19
a1	15.8	4.18	− 0.43	−0.33	− 0.38
a2	15.8	17.89	− 1.53	−1.00	− 1.27
a3	15.8	58.31	− 5.50	−3.35	− 4.44
b0	18.3	0.63	− 0.25	−0.22	− 0.23
b1	18.3	3.29	− 0.71	−0.52	− 0.61
b2	18.3	16.58	− 3.47	−2.24	− 2.86
b3	18.3	54.02	−12.18	−7.63	− 9.93
c0	22.9	0.58	− 0.28	−0.23	− 0.25
c1	22.9	3.64	− 1.00	−0.64	− 0.82
c2	22.9	15.32	− 4.22	−2.42	− 3.33
c3	22.9	49.98	−14.45	−8.17	−11.34

Although more realistic scenarios for the assumed emissions are used here than in the prior Johnston, et al. sensitivity analyses, the main conclusion is similar: the largest effects on total column ozone occur with the largest NO_x emissions and the highest assumed flight altitude. The Northern Hemisphere has the largest decreases in total ozone. Figure 5 shows the change in globally and annually averaged total ozone as a function of the altitude and magnitude of the emissions. This figure indicates that, for these scenarios, the assumed HSCT fleet would need to keep NO_x emissions below 0.5-1 x 10^{34} molecules per year, depending on the aircraft flight altitude, in order to affect globally-averaged total ozone by one percent or less. Using a one percent criteria in this discussion is entirely arbitrary, since humanity has yet to determine what is an acceptable level of ozone destruction from such emissions. Figure 6 shows that the change in globally and annually averaged total ozone varies almost linearly with the magnitude of the NO_x emissions, assuming that the altitude and latitude variations in the emissions remain unchanged.

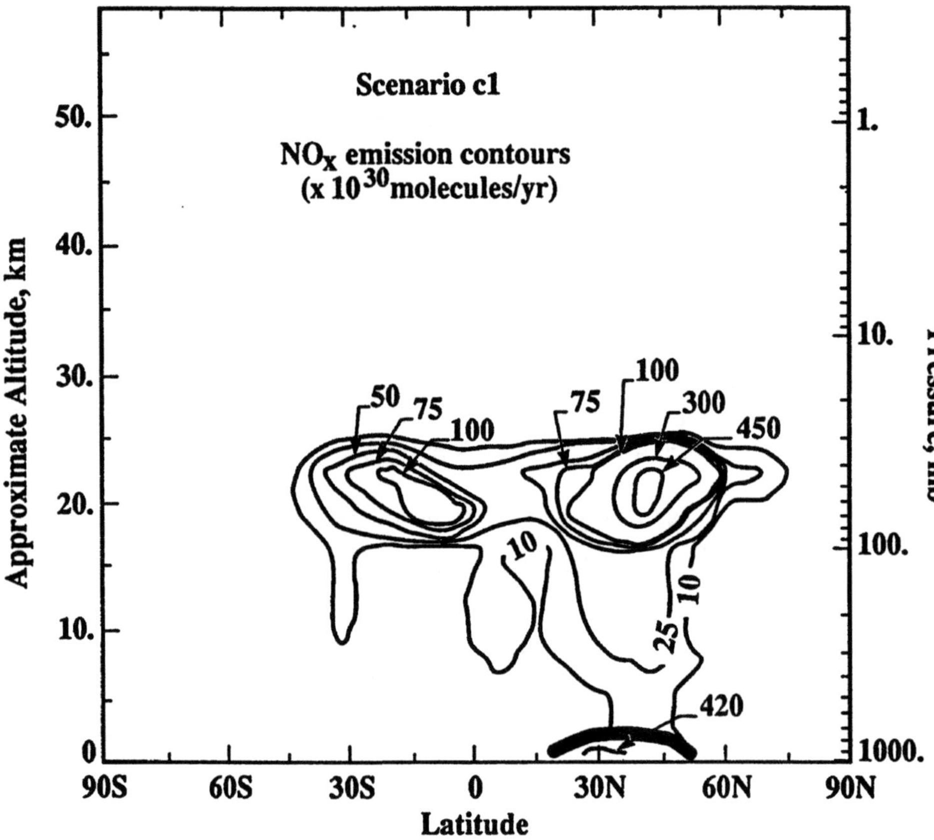

Figure 4. Emissions of NO_x as a function of altitude and latitude assumed in scenario c1, corresponding to an aircraft flying at a mean cruise altitude of 22.3 km and a total emissions of 3.64 x 10^{33} molecules of NO_x per year.

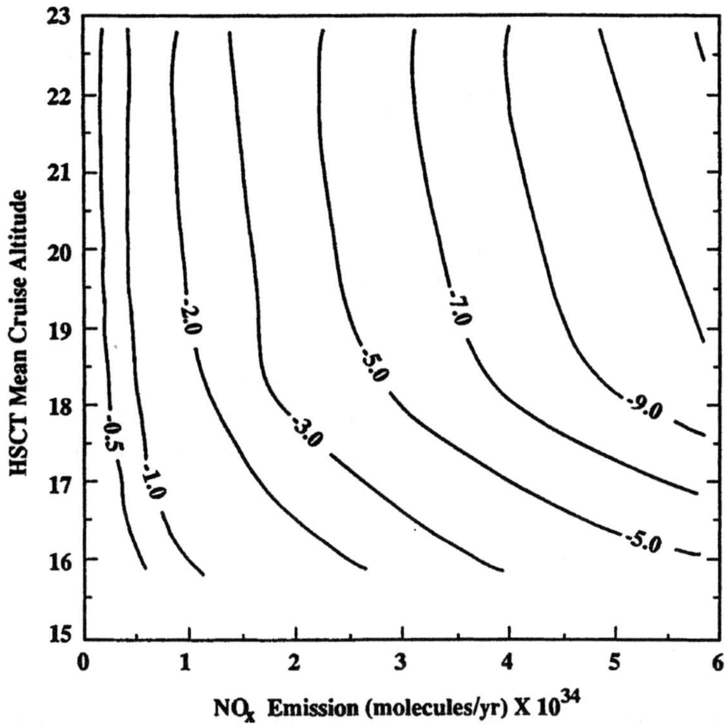

Figure 5. Change in globally and annually averaged total ozone at steady-state plotted as a function of mean cruise altitude and amount of NO_x emissions assumed in the twelve scenario matrix.

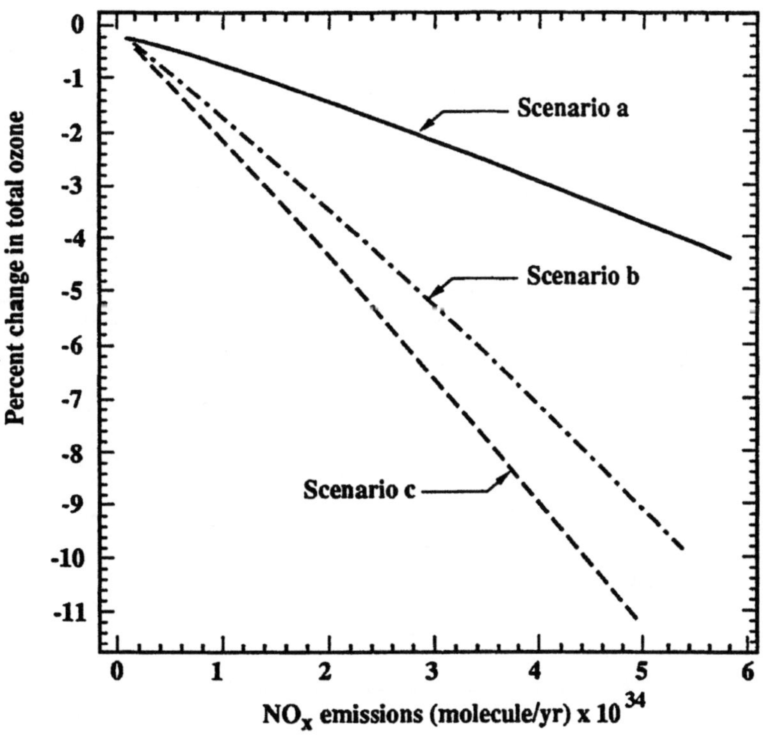

Figure 6. Change in globally and annually averaged total ozone as a function of NO_x emissions for aircraft scenarios with mean cruise altitudes of 15.9, 18.3, and 22.3.

Figure 7a gives the change in total ozone versus time of year for scenario c1, while Figure 7b shows the change in ozone versus altitude and latitude for the same scenario. The maximum change in total ozone occurs at high Northern Hemisphere latitudes and peaks in late summer and into fall. The minimum decrease in total ozone occurs in the tropics. Figure 7b shows that the maximum ozone destruction for this scenario occurs near 20 km altitude in Northern Hemisphere polar region.

Part of the intention in undertaking this analysis is to develop the appropriate criteria needed by the aircraft industry to ensure that the HSCT developed is not environmentally harmful. The U.S. aircraft companies are well-aware of the concerns about ozone and, through NASA's HSCT research program, are aware of our research. Because this matrix of scenarios extends over such a wide range of emissions we want to be careful that the results for these scenarios are not over-interpreted. We do not want to imply that the large emission cases in any way correspond to any fleet under actual consideration. In fact, our results suggest that the larger emission cases will need to be avoided in considering fleet designs.

OZONE RECOVERY TIME

Of interest to policymakers is the question of how long it takes for the ozone concentrations to recover if the emissions are greatly reduced. For scenarios a2, b2, and c2, Figure 8 shows that it takes about nine years after emissions begin for the maximum change in total ozone to be reached, and about an additional decade for the globally-averaged ozone column to recover to unperturbed conditions after emissions cease. The altitude of the emissions appears to make little difference in the recovery time.

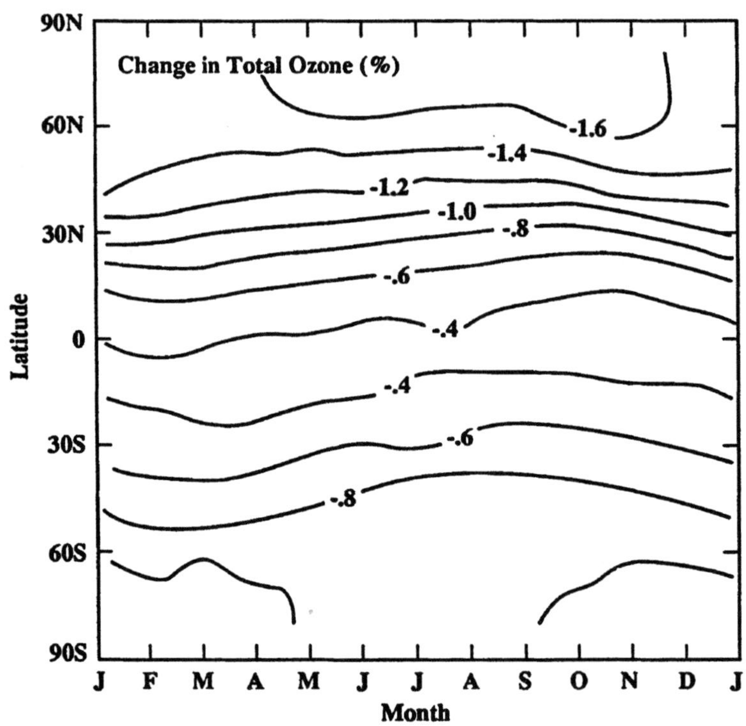

Figure 7(a)

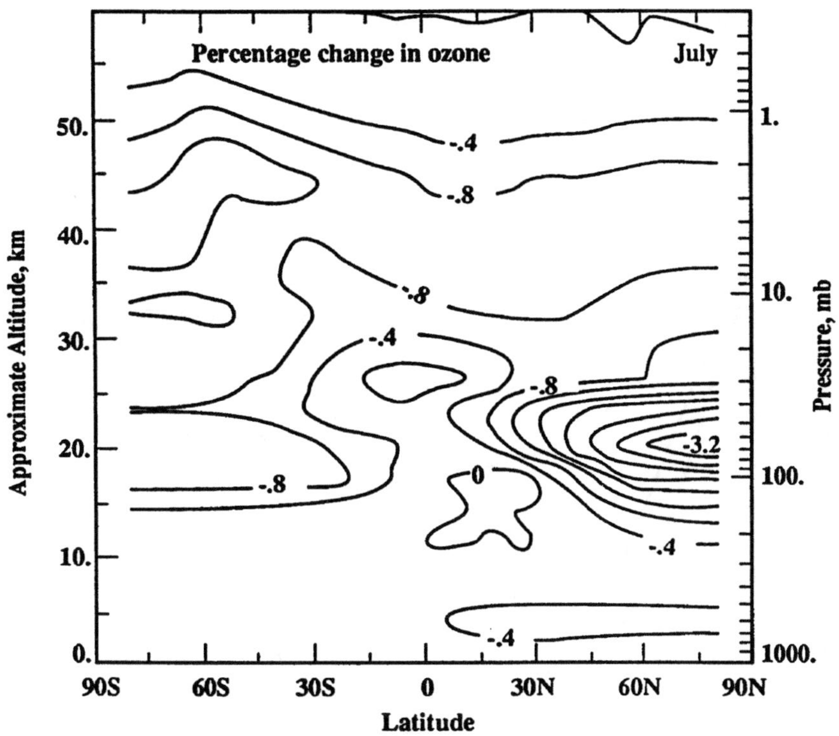

Figure 7(b)

Figure 7. (a) Calculated percentage change in total ozone at steady-state plotted as a function of latitude and time of the year as determined for scenario c1; (b) Calculated percentage change in ozone at steady-state plotted as a function of altitude and latitude as determined for scenario c1 during the month of July.

FURTHER SENSITIVITY STUDIES

This evaluation of the future HSCT fleet scenarios has, until this section, considered the effects on ozone resulting from NO_x emissions only. Several sensitivity calculations are evaluated here based on assumed emissions of water vapor and carbon monoxide. These estimated emissions are based on current engine technology. An additional sensitivity calculation will consider the effect on the calculated ozone changes of a future atmosphere with higher background stratospheric chlorine along with increased concentrations of CO_2, CH_4, and N_2O.

The stratospheric water vapor emissions due to aircraft emissions for hydrocarbon fuel combustion are much larger than the NO_x emissions. For scenario c1, this corresponds to about a factor of 370 larger number of molecules of water emitted into the stratosphere than NO_x emitted. The effect of the water vapor emissions by itself is small, but it has a larger impact when included in combination with the NO_x emissions. The effect on ozone with the coupled NO_x and H_2O emissions case is a factor of 0.9 less than the NO_x emissions only case. This effect is due the interactions between NO_x and HO_x chemistry.

The effects of carbon monoxide emissions were also evaluated. For scenario c1, the total CO emissions are 6.24×10^{33} molecules per year, but over half of these emissions occur in the bottom 1.5 km of the atmosphere. We found that the CO emissions had a negligible effect on global ozone.

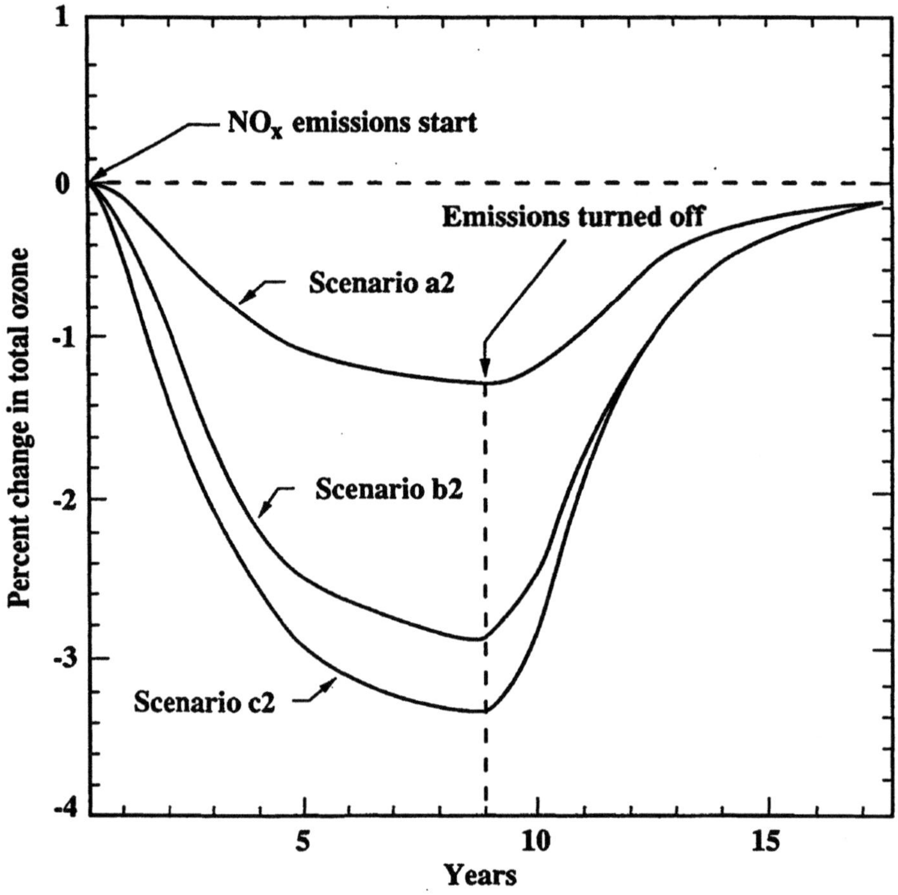

Figure 8. Change in globally and annually averaged total ozone as a function of time after initiating the NO_x emissions for scenarios a2, b2, and c2, followed by cessation of all emissions once steady-state has been reached.

As a test of the sensitivity of the aircraft scenario effects to the background atmosphere, we evaluated the effect of the NO_x emissions in scenario c1 for an atmosphere containing 5.2 ppbv of stratospheric reactive chlorine, 395 ppmv of carbon dioxide, 2.25 ppmv of methane, and 0.334 ppmv of nitrous oxide. This corresponds to the atmosphere in 2020 if one assumes constant emissions of CFCs at 1986 levels, and continuing trends at current rates for CO_2, CH_4 and N_2O. In actuality, this level of chlorine will likely not be reached in the next few decades as a result of international efforts to reduce CFC production. However, this chlorine level is useful in evaluating the sensitivity of aircraft emissions to background chlorine levels. There appears to be a large sensitivity to the background atmosphere; the calculation of scenario c1 with the 2020 background atmosphere determines a global total ozone decrease of –0.58% as compared to the ozone decrease of 0.82% calculated using the current atmosphere as the background. The chemical interactions of NO_x with stratospheric chlorine are likely responsible for the sensitivity, but the role of the increased background CO_2, N_2O, and CH_4 require further evaluation.

CONCLUSIONS

This study has used the LLNL two-dimensional model of the global atmosphere in an evaluation of the effects on global ozone concentrations from current subsonic aircraft emissions and from the emissions of possible future high speed civil transports. We have attempted to include more realistic representations of emissions as a function of altitude and latitude in these scenarios than were included in previous sensitivity analyses. Major findings from this study are:

1. Current aircraft emissions may be having an impact on upper tropospheric ozone, leading to increasing concentrations of ozone in the upper troposphere. However, the analysis here was very preliminary. The potential significance of the effects on ozone warrant much further study.

2. A matrix of HSCT scenarios evaluated over a wide range of mean flight altitudes and magnitudes of NO_x emissions confirmed previous analyses showing that ozone destruction becomes larger as the emissions of NO_x increase and as the altitude of injection increases.

3. Model calculations indicate that a major reduction in emissions would allow the stratosphere to recover to unperturbed conditions in about a decade.

4. Sensitivity studies indicate that water vapor emissions have a moderate effect on the change in total ozone, while carbon monoxide emissions had a negligible effect. Water vapor emissions should be considered in all future evaluations of HSCT fleets.

5. The calculated change in ozone for the HSCT scenarios was very sensitive to the background atmosphere, particularly to the levels of stratospheric chlorine and concentrations of carbon dioxide, methane, and nitrous oxide. This requires further evaluation, but suggests that assessment analyses of future HSCT fleets should carefully consider the appropriate background atmosphere for the time period in which the aircraft will fly. These analyses should also consider projections of subsonic emissions.

ACKNOWLEDGMENTS

Work at Lawrence Livermore National Laboratory was performed under the auspices of the U.S. Department of Energy under Contract W-7405-ENG-48 and was supported in part by NASA's Upper Atmospheric Research Program.

REFERENCES

Bauer, E., 1978: A catalog of perturbing influences on stratospheric ozone, 1955–1975, U.S. Department of Transportation report no. FAA-EQ-78-20.

Boeing Commercial Airplanes, 1989: High-speed civil transport study, National Aeronautics and Space Administration Contractor Report 4233.

Climatic Impact Assessment Program, 1974: Report of findings: The effects of stratospheric pollution by aircraft, edited by A. J. Grobecker, S. C. Coroniti, and R. H. Cannon, Jr., U.S. Department of Transportation report DOT-TST-75-50, Washington, D.C.

Climatic Impact Assessment Program, 1975a: Propulsion effluent in the stratosphere, Monograph 2, U.S. Department of Transportation report DOT-TST-75-52, Washington, D.C.

Climatic Impact Assessment Program, 1975b: The stratosphere perturbed by propulsion effluent, Monograph 3, U.S. Department of Transportation report DOT-TST-75-53, Washington, D.C.

COMESA, 1976: The report of the committee on meteorological effects of stratospheric aircraft, Parts 1 and 2, United Kingdom Meteorological Office, Bracknell.

Comite d'Etudes sur les Consequences des Vols Stratospheriques (COVOS), 1976: Activites 1972–1976, Societe Meteorologique de France, Boulogne.

Harrison, H., 1970: Stratospheric ozone with added water vapor: Influence of high-altitude aircraft, *Science*, **170**, 734–736.

Johnston, H. S., 1971: Reduction of stratospheric ozone by nitrogen oxide catalysts from SST exhaust, *Science*, **173**, 517–522.

Johnston, H. S., D. E. Kinnison, and D. J. Wuebbles, 1989: Nitrogen oxides from high-altitude aircraft: An update of potential effects on ozone, *J. Geophys. Res.*, **94**, 16351–16363.

Kinnison, D. E., 1989: Effect of trace gases on global atmospheric chemical and physical processes, Ph.D. thesis, University of California, Berkeley.

Kinnison, D. E., D. J. Wuebbles, and H. S. Johnston, 1988: A study of the sensitivity of stratospheric ozone to hypersonic aircraft emissions, Proceedings of the First International Conference on Hypersonic Flight in the 21st Century, Grand Forks, N. D., Sept. 20–23.

Kinnison, D. E., and D. J. Wuebbles, 1989: Preventing depletion of stratospheric ozone—implications on future aircraft emissions, Air and Waste Management Association paper 89-4.7; also Lawrence Livermore National Laboratory report UCRL-99926.

National Academy of Sciences (NAS), 1975: *Environmental Impact of Stratospheric Flight*, Washington, D.C.

World Meteorological Organization (WMO), 1985: Atmospheric Ozone 1985: Assessment of our understanding of the processes controlling its present distribution and change, Global Ozone Research and Monitoring Project–Report No. 16.

Wuebbles, D. J., 1983: A theoretical analysis of the past variations in global atmospheric composition and temperature structure, Ph.D. thesis, University of California, Davis; also Lawrence Livermore National Laboratory report UCRL-53423.

Wuebbles, D. J., and D. E. Kinnison, 1989: A two-dimensional model study of past trends in global ozone, in *Ozone in the Atmosphere*, edited by R. D. Bojkov and P. Fabian, A. Deepak Publishing, Hampton, Va.

DISCLAIMER

POSSIBLE CLIMATIC EFFECTS OF CONTRAILS AND ADDITIONAL WATER VAPOUR

H. Grassl

Max-Planck-Institut für Meteorologie, Hamburg
Meteorologisches Institut der Universität Hamburg

ABSTRACT

The importance of contrails in the upper troposphere and of additional water vapour in the lower stratosphere, both the result of aircraft emissions, for the radiation budget of the surface/atmosphere system is roughly assessed. The radiation flux density profiles with contrails and additional water vapour are compared to other greenhouse gas forcings. This leads to a very first order of magnitude estimate of the air traffic climate forcing potential: two percent contrail cover or air traffic induced natural cirrus may be as important for the planetary radiation budget as a 10 percent increase of present anthropogenic CO_2 forcing, equivalent to six years emission of 25 Gigatons CO_2 per year; additional water vapour in the lowest high northern latitude stratosphere is considerably contributing to the greenhouse effect of the atmosphere. If air traffic would cause a 10 percent increase in water vapour there this would be equivalent to up to 0.2 Wm^{-2} radiation budget change depending on surface temperature.

INTRODUCTION

Although the Club of Rome warned nearly two decades ago of a shortage of resources we are in the contrary now mainly discussing pollution problems. Here we have to ask whether air traffic is contributing considerably to the following major air pollution effects:

- increase of the greenhouse effect of the atmosphere provoking a global warming
- observed depletion of the stratospheric ozone layer leading to a thinning of the shield against ultraviolett radiation in the UV-B range from 0.28 to 0.32 micrometer wavelength

- increased occurrence and severity of photochemical smog in the troposphere mainly caused by NO_x and hydrocarbon emissions increasing also average tropospheric ozone levels
- Widespread increased acidity of rain acidifying lakes and soils and causing corrosion of buildings
- eutrophication of marginal seas by wet and dry nitrate deposition, whereby the nitrate is the product of chemical reaction involving NO_x as precursor gas.

Since air traffic is a source of carbon dioxide (CO_2), nitrogen oxides ($NO_x = NO + NO_2$), carbon monoxide (CO), a variety of hydrocarbons (HC) and particles (often soot) it must contribute to all of them except ozone depletion in the upper stratosphere. However, this exception is only due to cruising levels mostly below 13 km, still too low for this effect (if also accounting for the general downward motion in the high latitude lower stratosphere). The main question is: How and where is the rather small air traffic contribution to total global emission enhanced? We have, in other words, to explain why water vapour emission by aircraft engines is a problem, although a lake of 150 km^2 evaporates roughly the same amount as is injected by the global fleet of aircraft.

The following pages will be restricted to the water vapour emission only. After a section 2 on the basic atmospheric temperature structure and water vapour abundance contrail formation as a function of pressure and temperature is introduced in section 3. The radiation flux density changes caused by thin cirrus (here taken as represen-tative for contrails, which may not be correct in all cases) are discussed in section 4 including a first order of magnitude estimate of contrail impact on the radiation budget. Section 5 then discusses water vapour in the lower stratosphere from the point of view of radiation flux density changes. A final section 6 concludes with a summary of findings and open questions.

ATMOSPHERIC STRUCTURE AND RESIDENCE TIME OF AN ADMIXTURE

The density of well mixed gases like oxygen (O_2) and carbon dioxide (CO_2) decreases like pressure exponentially with height halving the value approximately every 5.5 kilometers. Thus emission in 11 km at 250 hPa, a fourth of surface pressure, is for all substances a stronger disturbance than at lower levels even if accounting for a reduction in fuel consumption due to decreased air resistance. If the substance in question is falling off more rapidly with height than pressure, for instance water vapour, then aircraft emissions become more and more important, might even become dominant for the concentration of a trace substance from a certain height level upwards. This is illustrated in Table 1 and Figure 1. Temperature decreases typically by 0.6 K per 100 m ascent in the troposphere and water vapour saturation pressure is for T < 10°C at least halved by a 10 K temperature decrease. The result: water vapour is a trace gas in the parts per million by volume range at present subsonic jet engine cruising levels.

Table 1: Water vapour pressure at saturation.

Temperature	saturation pressure over water	saturation pressure over ice
30°C	42.42 hPa (*)	
20	23.94	
10	12.27	
0	6.10	6.10
- 0	2.86	2.59
- 20	1.25	1.01
- 30	0.50	0.37
- 40	0.19 *	0.12
- 50		0.039
- 60		0.010
- 70		0.0026
- 80		0.00054
- 90		0.00009
- 100		0.00001

* Supercooled water droplets often exist down to -40°C.
(*) Volume mixing ratios are calculated by dividing by total pressure in the level of observation.

Another basic feature of our atmosphere, which is of relevance here, is the higher tropopause over warm surfaces. Therefore, nearly all present subsonic aircraft never reach the tropical stratosphere, experience coldest temperatures during advection of subtropical air, when the tropopause is roughly at 13 km.

The potential atmospheric residence time of an admixture "jumps" from a few weeks for the upper troposphere to typically a year or more in the lower stratosphere. Residence time is meant as the time a molecule or particle - if not undergoing chemical transformation - typically resides in the respective part of the atmosphere before entering another part or being mixed down to the surface where it is eventually deposited.

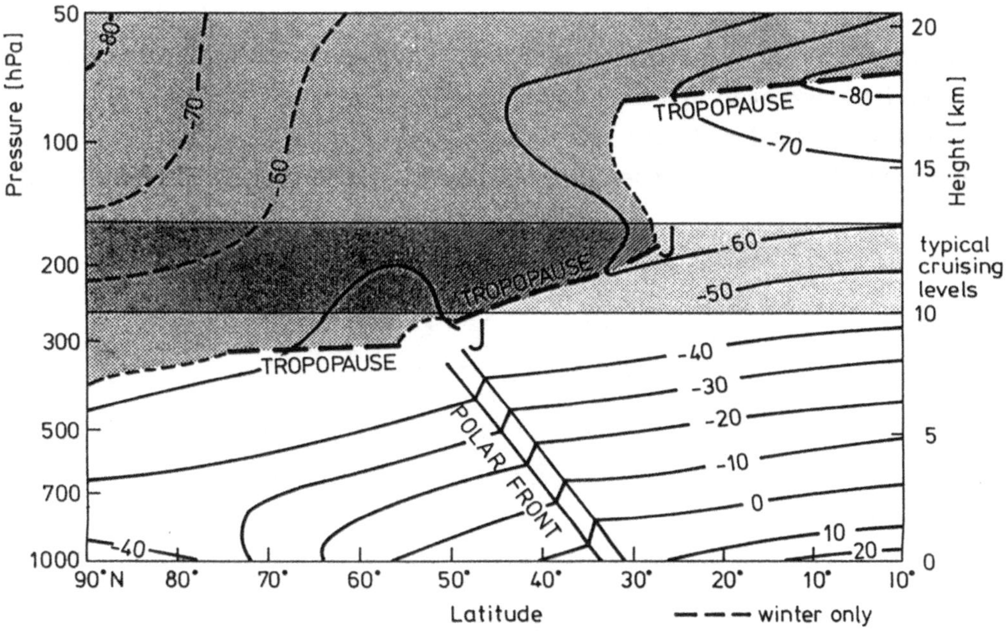

Fig. 1: Temperature structure of the Northern Hemisphere winter atmosphere, shown in a latitude-height isoline plot (°C) with indication of mean tropopause level and typical subsonic jet aircraft cruising levels. Dashed lines in the stratosphere show minimum temperatures rather than averages.

CONTRAIL FORMATION

Contrails (condensation trails) may form in the exhaust plume of a jet engine if a volume of air is supersaturated with respect to a plane water or ice surface. Supersaturation in this case is reached by mixing of two unsaturated air volumes of different temperatures, here by cooling the engine exhaust containing the additional water vapour and many condensation nuclei through dilution with ambient air. Since the saturation vapour pressures (see also table 1) is lower over ice, the contrail, starting as a water cloud in the still warm exhaust air stream, is after a short time easily transformed into an ice cloud because the temperatures are low enough (below -40°C) for icing of any cloud, even without any ice nuclei. Thus contrails lasting for more than a few seconds show a spectral transmission typical for ice clouds (Grassl, 1970). Already in 1953 Appleman has calculated the temperature and pressure values leading to contrail formation if kerosene is burnt. Fig. 2 repeats the basic features and indicates, by including an observed and a mean temperature profile, that present cruising levels very often tend to optimize inadvertently contrail formation. Roughly speaking flying in 12 km at -60°C has to cause a contrail, whose lifetime is proportional to the distance between actual temperature and the 0% relative humidity line in Fig. 2, hence under these conditions the aircraft emitts enough water vapour to cause an ice cloud even in a dry atmosphere. The contrail lifetime is also dependent on lateral and vertical diffusion which is not only influenced by the dynamical and thermal structure of the atmosphere but also by aircraft and engine design. Despite

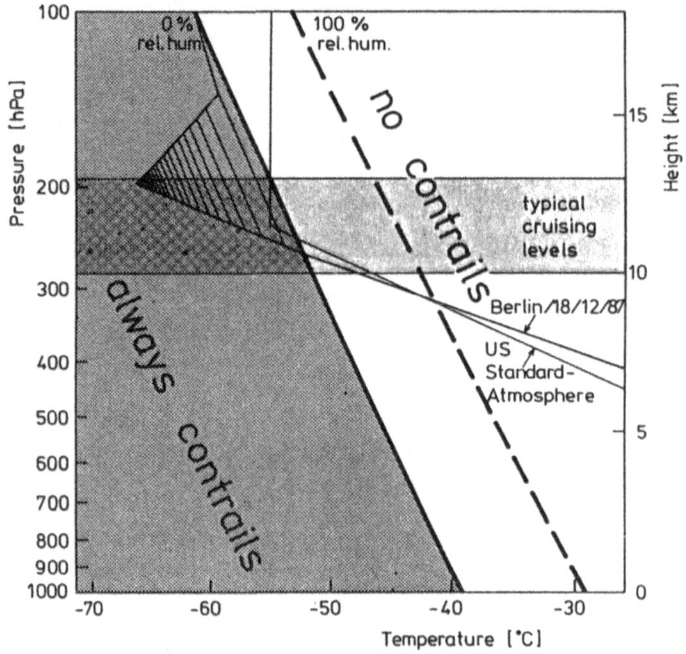

Fig. 2: Contrail formation depending on air pressure, air temperature and relative humidity, also indicating main cruising levels, a standard temperature profile as well as an individual one; modified after Appleman (1953).

these many influences Fig. 2 possibly points to a way for the assessment of the contrail abundance. Observation of the lifetime for several cases for a distinct aircraft could allow the extrapolation to all radiosoundings of a distinct station and might even lead to a contrail lifetime forecast for a distinct meteorological situation. The main drawback of this proposal is the poor quality of the water vapour measurement by radiosondes at temperatures below about -30°C.

A simple graphical representation of a simple calculation of the relative humidity increase by an airplane at cruising altitude should underline the importance of temperature (Fig. 3). Let us suppose we fly between 9 and 12 km height and the aircraft burns 1 ton kerosene per 100 km, i.e. emits 1.25 tons water vapour. This water vapour is diluted into a 1 km wide and 150 m thick exhaust channel. Then we would raise for example relative humidity and thus water vapour concentration by 6 percent at 10.5 km and -60°C.

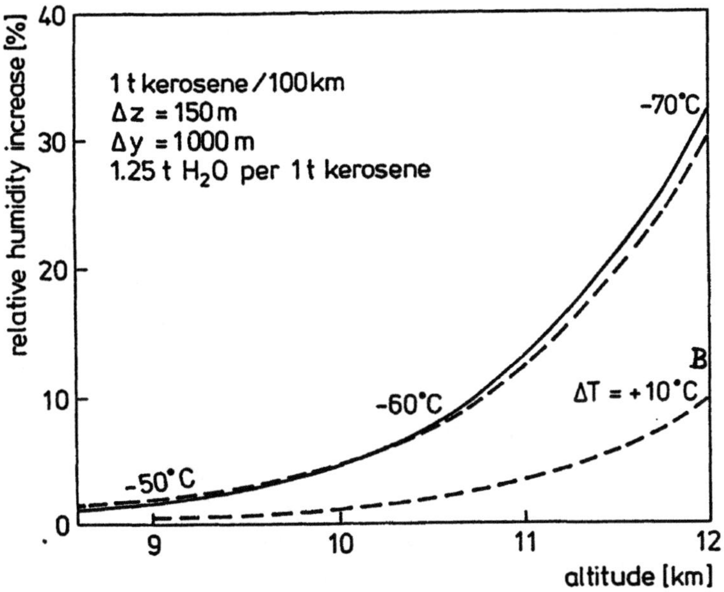

Fig. 3: Relative humidity increase as a function of altitude and corresponding temperature; the dashed line includes increased efficiency at higher cruising altitudes. The lower dashed curve is valid for higher temperature (+ 10 K).

The temperature height curve assumed is quite realistic as indicated by the Berlin radiosonde profile in Fig. 2, where temperature drops by 15K from 10 to 12 km. The corresponding maximum relative humidity increase at 12 km is plotted as B in Fig. 3.

RADIATIVE FLUX DENSITY CHANGES BY THIN ICE CLOUDS

High clouds have generally a strong impact on thermal radiation transfer in the atmosphere (Liou, 1986), since they absorb heat radiation from the warm surface in otherwise transparent spectral intervals of a cloudless atmosphere and reradiate far less upward as well as downward, dictated by their low temperatures. Hence, they effectively shield the planet from a higher heat loss to space. This shielding may be offset during daytime by reflecting back to space solar radiation. The thermal radiation change is best understood by discussing the radiation flux density profiles for different thin ice clouds. Fig. 4 clearly points to the drastic reduction of radiation to space (here only shown up to 20 km height) by optically thin cirrus. Approximately 100 Wm^{-2} less heat radiation is emitted to space if an ice cloud with an optical depth of 1.8 (this would be a very strong contrail) interferes. The crystal size distribution assumed has been proposed for cirrostratus clouds. Since the thermal net flux density F_{net} is strongly decreasing with increasing height in the lower portion of the cirrus layer this

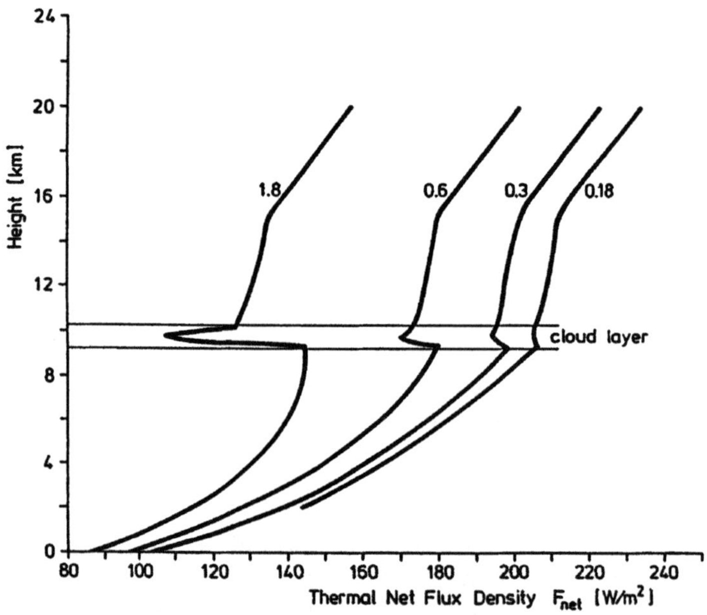

Fig. 4: Thermal net radiation flux density profiles for different thin ice clouds of 750 m vertical extent but varying optical depth as indicated; modified after Manschke, 1985.

layer is heated through absorption of radiation from below. Only the upper portion with a strong increase of F_{net} with height shows a cooling. One can imagine that radiation processes acting long enough in this way tend to destabilize the temperature profile causing turbulent mixing and a contrail cross section change. This radiation forced vertical mixing is well known to all of us as the bumpiness when flying near to a cloud top. Vertical net flux density gradients may be converted into instantaneous heating rates by

$$\frac{\partial T}{\partial t} = - \frac{1}{\rho c_p} \frac{\partial F_{net}}{\partial z}$$

with ρ = air density, c_p = specific heat of air at constant pressure.

At 250 hPa with air density ρ of roughly 0.35 kg m^{-3} and 20 Wm^{-2} net flux difference within 300 m cooling or heating depending on the sign of the difference already amounts to ~1K per hour. Most important for the following discussion of possible contrail effects is the F_{net} change for very thin ice clouds. A cloud with an optical depth δ = 0.1 already reduces net thermal radiation to space by about 10 Wm^{-2}, depending on height and temperature as well as microphysical properties of ice clouds. This exceptionally strong radiation budget change is due to the following physical facts:

- very low temperatures near the tropopause lowering emission of an absorber
- strong absorption by ice crystals in the entire thermal infrared
- crystal sizes large enough to cause a relative extinction efficiency of approximately 2, but still small enough to lead to high optical depth

  $$\delta = 2 \int \Pi r^2 \, n(r) \, dr$$

  with r = cross section equivalent radius of a cirrus crystal,
  n(r) = crystals per volume per radius unit.
  It has been shown by Kinne (1982) that the cross section equivalent radius is a better approximation than surface or volume equivalent radius.

Thin cirrus clouds therefore are most effective in shielding the Earth from a heat loss.

The foregoing discussion did not account for backscattering of solar radiation by these clouds. If included, a more complicated picture emerges. Fig. 5, also like Fig. 4 taken from a master's thesis of a former student, demonstrates that optically thin ice clouds increase the greenhouse effect of the atmosphere in most cases, because the solar net flux density change $\Delta F_{net,\,s}$ at the top of the atmosphere is smaller than ΔF_{net} in the thermal infrared. Only for very low sun elevation $\Delta F_{net,\,s}$ surmounts ΔF_{net}. One should note, however, that all curves would at all sun elevations be lower for lower optical depth, typical for most contrails. Taking $\Delta F_{net,\,s}/\Delta F_{net}$ = 0.5 (no exaggeration, rather a conservative estimate) for δ = 0.1, we have to compare a 5 Wm^{-2} net radiative budget forcing with other anthropogenic forcings. Doubling CO_2 for example gives $\Delta F_{net} \cong$ - 3 Wm^{-2} at the top of the atmosphere at fixed other parameters, as also assumed for the calculations with thin ice clouds. If such a radiative forcing is applied to a radiative-convective equilibrium climate model the resulting forcing of the troposphere/surface system expressed in a net flux change at the tropopause is ΔF_{net} = 4 Wm^{-2}. Radiative-convective equilibrium in this context means that the radiation balance of the Earth has been reached again by heating of the lower atmosphere and the surface, thereby allowing for convection in the troposphere. Since industrialisation began, CO_2 content increased by 25 percent, the socalled CO_2-greenhouse forcing of the troposphere/surface system therefore ≅ - 1 Wm^{-2} at present. A two percent cover with such contrails would therefore be equivalent to 10 percent of the present CO_2 greenhouse forcing or to roughly 5 years of CO_2 emissions with presently 25 Gigatons CO_2 per year. This rough first estimate should not be misinterpreted, it is simply an indication that adding water vapour into very cold layers leading to contrails acts as a strong amplifier of radiation budget changes. While air traffic contribution to CO_2 emissions is in the 1-2% range, and emission height is irrelevant because of the long lifetime, high tropospheric water vapour injection strongly amplifies the environmental impact of air traffic. At the same time this underlines the difficulties climate modellers have in correctly describing the role of ice clouds in a warmer 2 x CO_2 climate.

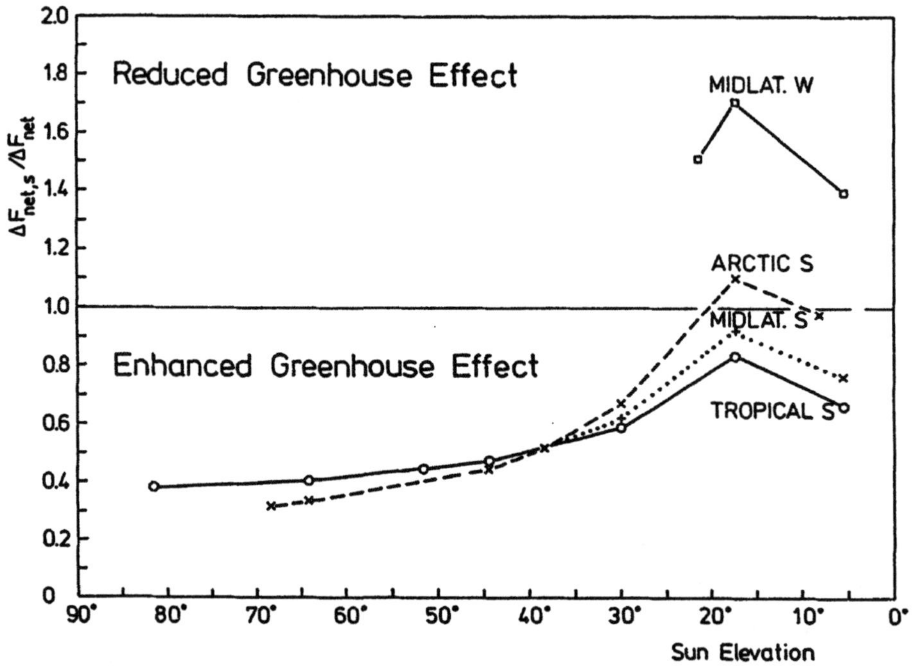

Fig. 5: Net effect of ice clouds (1 km cirrostratus with $\delta = 1$) on the radiation budget shown as the ratio of solar radiation change to thermal radiation change ($\Delta F_{net,\,s} / \Delta F_{net}$) as a function of sun elevation; for four different standard atmo-spheres as indicated; modified after Kinne (1982).

Frequently the following argument is put forward: High level cirrus is of nearly no influence if above an optically thick lower cloud deck. With respect to contrails this would mean: only contrails in otherwise clear skies have a strong radiation budget change capacity. However, this is not true, since the backscattering of solar radiation is reduced by a thin cirrus above an optically thick water cloud (Bumke, 1984) and the shielding of thermal radiation is mainly a function of the temperature difference between lower cloud deck and cirrus clouds. Over low lying stratus decks contrail impact is thus enhanced if compared to a cloudfree condition.

The above calculations of radiation flux density changes ΔF_{net} and $\Delta F_{net, s}$ applied measured crystal size spectra of cirrostratus, a natural ice cloud. These crystals are rather large (150 μm length, 30-40 μm width) and there are good arguments that crystals in contrails are far smaller owing to the evolution of contrails. They start as water clouds and are rapidly transformed into ice clouds. Since all known shortlived water cloud droplet size spectra peak in the 2 to 10 μm radius range, a better representation would be to use strongly smaller ice crystal sizes. However, this is only a physically plausible proposal not at all underlined by measurements, which do not exist, because sizing instruments typically stop at 50 μm diameter crystals. An additional argument for small crystal sizes is the increase in condensation nuclei numbers in the engine exhaust again leading to smaller typical droplet sizes freezing at -40°C the latest. An increase in crystal numbers at constant ice content in a contrail strengthens solar radiation back-scattering at a given ice content. Since we do not know the typical optical depths of long lived contrails the net effect in both radiation domains, solar and terrestrial, cannot be assessed more accurately. Two examples for a possible change: An increase of $\delta = 0.1$ to $\delta = 0.2$ would add to the greenhouse forcing, an increase from $\delta = 1$ to $\delta = 2$ would rather subtract.

WATER VAPOUR IN THE LOWER STRATOSPHERE

Water vapour is the most important radiatively active gas in the Earth's atmosphere and is responsible for more than two thirds of the greenhouse effect of the atmosphere. With its numerous absorption bands any concentration change in any layer affects the planetary radiation budget. As for other spatially varying gases changes in concentration are most effective at the tropopause level. Additionally accounting for the possibility of accumulation through increased residence time the lowest stratosphere is the most sensitive area for radiation flux density changes. In this context it is important to note that stratospheric water vapour content in undisturbed conditions is mainly controlled by two sources: Firstly, the source in anvils of tropical cumulonimbi penetrating a short distance into the stratosphere, where due to the very low temperatures around -80°C or even below the low water vapour volume mixing ratio of 2-3 ppm is determined; secondly, the source through oxidation of methane causing an increase in mixing ratio upwards and polewards from the tropical tropopause, to some-

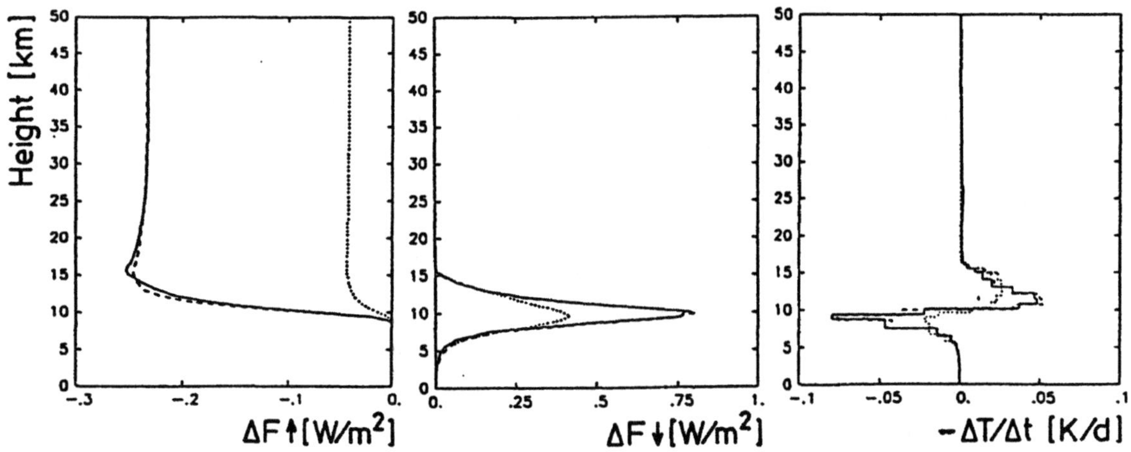

Fig. 6: Upward (a) and downward (b) thermal flux density change (ΔF↑, ΔF↓) caused by a 10 percent increase in water vapour density from 10-15 km for three atmospheres. Also changes in cooling rates (positive values mean stronger cooling) are shown in (c); (—) tropical atmosphere, (---) midlatitude summer atmosphere, (. . .) arctic winter atmosphere, after Hollweg (1990).

what higher values from 3-4 ppm on average. Again, like for contrails, air traffic is injecting water vapour into a rather sensitive area. In order to estimate the amount injected a simple calculation reveals: If from the yearly consumption of 120 million tons of fuel resulting in 150 million tons of water vapour 20 percent are injected into the stratosphere northward of 40°N (Lufthansa gave 15-17 percent for a test under summer conditions) this constitutes 15 percent of the typical water vapour content in the 10 to 13 km layer for the area north of 40°N. For this calculation a mixing ratio of 4 ppm has been assumed. Taking a residence time (here e-folding time) of one year a 5-6 percent increase in the 10-13 km layer would result. Higher flight levels would increase this percentage.

Does this affect the radiation budget? The answer from radiative transfer calculations in our institute, using a code tested by a line-by-linde code, is (Hollweg, 1990): A 10 percent increase of water vapour in the 10-15 km layer reduces emission to space by up to 0.23 Wm^{-2} depending on geographical area. The biggest ΔF_{net} occurs over warm surfaces. However, since the tropical tropopause is at 15-17 km height, the calculations represented by the full curve in Fig. 6 are valid for a 10% increase in upper tropospheric water vapour amount. The dashed curve for a midlatitude summer atmosphere with a tropopause at 11 km height realistically describes the effect of a 10% water vapour increase at the tropopause and lower stratosphere. The small decrease of F_{net} in polar winter atmospheres (dotted curve) is due to the cold lower troposphere. Taking into account the surface area of the Earth ($\sim 5 \cdot 10^{14}$ m^2) this 10% water vapour increase from 10-15 km is in terms of heat trapped equivalent to up to $1.2 \cdot 10^{14}$ W, more than ten times global anthropogenic energy throughput. It is also over warm surfaces comparable to a tenth of present mean greenhouse forcing amounting to ~2 Wm^{-2}. Figure 6 additionally points to a redistribution of heating or cooling rates by this additional water vapour. More cooling in the lower stratosphere and less cooling below 10 km, giving a tendency for an upward shift of the tropopause.

These model results indicate that already a few percent increase of lower stratospheric water vapour concentration are of considerable importance for the planetary radiation budget and for the cooling rate profile. This statement is a mere consequence of injection of water vapour into layers of most effective greenhouse forcing in the atmosphere.

DISCUSSION

Water vapour, by far the most important greenhouse gas of the atmosphere, is also a product of burning of kerosene by aircraft (1.25 tons per tone kerosene) and due to its injection into very cold layers around the tropopause of considerable environmental impact. Especially two effects have to be discussed with respect to present civil air traffic:

- contrail formation in the upper troposphere
- additional water vapour in the lowest stratosphere of high latitudes.

Both enhance either under most conditions (contrails) or always the greenhouse effect of the atmosphere. Most effective is water vapour injection at very low temperratures since the optical depth of contrails will be low enough for a greenhouse forcing, their lifetime will be increased and shielding of radiation to space is most pronounced if the absorber (contrails or water vapour) is cold. In our latitudes these low temperatures are mainly observed in subtropical airmasses when the tropopause height is above average.

The relative contribution to atmospheric composition increases with injection height in a threefold manner:

- density falls off exponentially, approximately halved every 5.5 km height increase
- maximum possible residence time of an admixture increases from days over weeks and months to years if ascending up to 13 km
- water vapour and nitrogen oxide (NO_x) background mixing ratios show a minimum in the lower stratosphere.

All the statements in this paper are qualitative in the sense that we lack the proof of a considerable increase in cloudiness through contrails and of water vapour concentration in the lower stratosphere in high northern latitudes. Attempts to identify contrail impact on local climate parameters where either not finding a significant modification (Rotter, 1987) or if finding an impact have also looked to parameters in the solar spectral range, where the impact is smaller than in the thermal range (Changnon, 1981). A first objective contrail cover estimate over Central Europe (Schumann, this volume) points to frequent occurrence. However, the method used can only give a minimum cover.

If one would for example find a rather low contrail cover this is, however, by no means an argument not to continue search for an air traffic water vapour and cirrus cloud impact, because the increase in high tropospheric water vapour content even without the existence of longlived contrails will be the cause of earlier "natural" cirrus cloud formation lateron at favorable meteorological conditions. This effect might even be stronger than that caused directly by contrails. Also the question of changed stratospheric chemistry through additional water vapour has not been discussed as well as the possible impact of H_2O and NO_x on wintertime polar stratospheric clouds. We are only now - with forecasts of a dramatic increase in air traffic worldwide - starting research on the effect of a water vapour injection. What is most urgently needed is measurements around the tropopause.

REFERENCES

Appleman, H., 1953: The formation of exhaust condensation trails by jet aircraft. *Bull. American Meteorological Society,* **34**, 14-20.

Bumke, K., 1984: Berechnung der lokalen planetaren Albedo und der Erwärmungsraten für eine vorgegebene Wolken- und Aerosolvergößerung im solaren Strahlungsbereich. *Thesis, University of Kiel.*

Changnon, S.-A., 1981: Midwestern cloud, sunshine and temperature trends since 1901: possible evidence of jet contrail effects. *J. of Applied Meteorology,* **20**, 496-508.

Grassl, H, 1970: Bestimmung der Größenverteilung von Wolkenelementen aus spektralen Transmissionsmessungen. *Beitr. Phys. Atmos.*, **43**, 255-284.

Hollweg, H.-D., 1990: private communication.

Kinne, St., 1982: Einfluß von Eiswolken auf den Strahlungshaushalt der Atmosphäre. *Thesis, University of Hamburg.*

Liou, K.-N., 1986: Influence of cirrus clouds on weather and climate processes: a global perspective. *Monthly Weather Review,* **114**, 1167-1199.

Manschke, A., 1985: Einfluß dünner Cirren auf den langwelligen Strahlungshaushalt der Atmosphäre. *Thesis, University of Kiel.*

Rotter, M., 1987: Auswirkung von Flugzeugkondensstreifen auf die Sonnenbeobachtung und die Bewölkungsverhältnisse in Kärnten. *Dissertation, Universität Graz.*

DETERMINATION OF CONTRAILS FROM SATELLITE DATA AND OBSERVATIONAL RESULTS

U. Schumann and P. Wendling

DLR, Institute of Atmospheric Physics,
D-8031 Oberpfaffenhofen, Germany

ABSTRACT

Condensation trails (contrails) from air traffic in the upper troposphere may have an effect on the earth's climate. We first discuss the state of knowledge. Then some results of radiation transfer calculations are reported which help to estimate the order of magnitude of the effects of cirrus clouds and water vapour in comparison to doubling of the concentration of CO_2. The main part of the paper describes the analysis of contrails from satellite data. Based on NOAA-AVHRR data we present results for a specific case where contrails covered a few percent of a region including southern Germany and the Alps. Preliminary results are reported for 99 further cases. A pattern recognition method is described to identify contrails and the related cloud cover automatically. Laser observations reflect contrail cross-sections which are several km in width and about 700 m thick. The optical thickness reaches values up to 1.

INTRODUCTION

It is known that thin cirrus clouds of large (more than about $3 \mu m$ in radius) ice particles in the upper troposphere at low and mid latitudes act to enhance the "greenhouse effect" owing to their rather high emissivity and low albedo (Stephens and Webster, 1981, Stephens, 1989). An increase in cirrus cloud coverage by a few percent might have the same effect as a doubling of the amount of CO_2 (Liou, 1986). According to recent results of global circulation models (Wetherald and Manabe, 1988), clouds (for fixed radiation parameters; Stephens, 1989) tend to enforce the greenhouse effect induced by increasing CO_2.

Contrails from aircrafts enhance the appearance of high-level clouds. Such contrails emerge in particular at temperatures below -40^{0}C (Appleman, 1953). Such low temperatures prevail at the tropopause of the atmosphere where the radiation effect of absorbing material is strongest. The altitude of the tropopause varies from typically 8 km at high latitudes to about 18 km in the tropics. Thus, flight levels at mid-latitudes reach or exceed the tropopause. In view of the strongly increasing long-distance air traffic (an increase by a factor of two is expected by air traffic experts for the next decade) it is important to investigate the effects of contrail-induced cirrus clouds on climate more closely. The need for such investigations (in addition to studies of changes in air chemistry by aircraft emissions) was stressed in several recent discussions (Held, 1988; Pfeiffer and Fischer, 1989).

Changnon (1981) reports about trends in cloudiness and sunshine based on routine surface observations for the time-period of 1901 to 1977 in the mid-west of the USA. He finds trends which seem to indicate an increase in cloudiness and decrease in sunshine duration in the period since 1960 and correlates this with the increasing fuel consumption by air traffic. Moreover, he finds a reduced temperature difference between day and night which might be caused by additional clouds. However, the data base for this analysis is too small to generalize such results. Angell (1990) reports on the cloud cover and sunshine duration at 100 stations within the whole USA. He compares mean values of the years 1950-68 with those in the years 1970-88 and shows that the cloud cover did increase by 2.0 ± 1.3 from the first to the second period. The sunshine duration, however, decreases by only $0.8 \pm 1.2\%$. He explains this with the stronger increase of thin cirrus clouds which are observed as additional cloud cover but are still too thin to reduce the sunshine-duration measurements. Weber (1989) finds also a decrease in sunshine duration at several stations in Germany, but he argues that this originates from an increase in westerly flow bringing more clouds from the Atlantic over the continent. From a study of contrails recorded by a sky-camera over a period of one year in Kärnten/Austria, Rotter (1987) observed long-living contrails with a persistency of more than one hour and a cloud cover of up to 6 % on nine days. For the majority of the observations, the cloud cover was well below 1 %.. Statistics on the frequency of contrails are also reported by Beckwith (1972). He found longliving contrails (more than 5 minutes persistency) in 25 % of all contrail observations from aircrafts at 13 km average flight level. Such observations are limited to a small region and confined to observations from the earth's surface or from aircrafts.

Satellite observations may provide objective measures to determine the cloud cover induced by contrails over a larger region. High resolution infrared satellite images often provide striking examples of contrails; corresponding visible images are usually less revealing (Joseph et al., 1975; Lee, 1989). Carleton and Lamb (1986) show that the occurrence of contrails can be determined from high-resolution (0.6 km in the visible and 1.0 km in the infrared spectral range) observations from the Defense Meteorological Satellite Program (DMSP). From a pilot study they find that contrails tend to occur frequently and often in association with natural cirrus clouds. Lee (1989) shows that data from the NOAA Advanced Very High Resolution Radiometer (AVHRR) can be used to produce images which greatly enhance contrails in comparison to the mono-spectral observations. He gives one example with little additional evidence on the meteorological conditions.

Except for in-situ measurements by Knollenberg (1972), little is known about the spatial structure and micro-physical parameters of contrails.

In this paper, we first discuss some estimates of the effects of water vapour and cirrus clouds near the tropopause on the heating rate of the atmosphere. Then results for a few cases are reported, mainly for the selected day of Oct. 18, 1989. For this day, ground observations, satellite data and Lidar observations will be presented. In particular, a new pattern recognition method will be described. The present study reports work in progress. Much more effort is required before we can make final recommendations with respect to aspects of air traffic.

RADIATION EFFECTS OF WATER VAPOUR AND CLOUDS

As stated before, thin cirrus clouds at higher levels have a strong effect on radiative heating. Fig. 1, taken from Liou (1986), shows results from a one-dimensional radiative transport model. The code is applied to compute infrared cooling (lower diagram) and solar heating (upper diagrams) rates for cloudy atmospheres using standard atmospheric temperature, water vapour and ozone profiles. Cirrus clouds with thicknesses of 0.1, 1, and 3 km are inserted in the atmosphere and a 100 % cloud cover is assumed. The base of the cloud is fixed at 8 km. For further parameters see Liou (1986). We see from the results that a cirrus cloud causes moderate heating by absorption in the solar range. The IR cooling rate reaches large values with a large cooling rate at the cloud top and a significant heating rate at the cloud base.

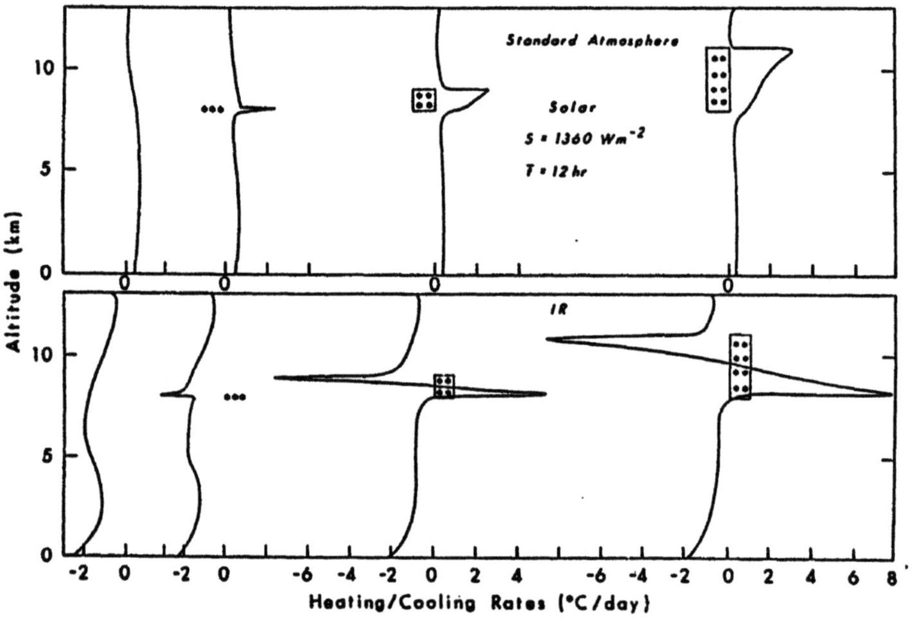

Figure 1. Thermal infrared and solar heating rates for cirrus clouds, taken from Liou (1986).

It should be noted that these are pure radiation calculations which do not reflect vertical transport processes induced in the atmosphere. Hence, the rather small effects of the clouds on the heating rate at the surface should not be misinterpreted. The vertical profile of cooling and heating will destabilize the layer taken by the cloud and thus induce additional vertical mixing, possibly with the effect of an increase in the altitude of the tropopause. Such an increase was found by Wetherald and Manabe (1988) and used to explain why high clouds in general tend to enhance the "greenhouse effect".

We have performed similar (but even simpler) radiation calculations with respect to the effects of increased water vapour. For the purpose of qualitative estimates, we investigate the change in infrared heating rate due to an increase of water vapour concentration (without clouds) in the altitude range from 9.5 to 14.5 km. We simply compare the results from two radiation calculations, one with the standard atmosphere and one with the water vapour increased up to local saturation with respect to water in the same atmosphere. The effect of changed water vapour concentration in the solar spectral range is negligible in comparison to that in the infrared. From Fig. 2 we see that the induced changes in heating rate due to changes in water vapour concentration (i.e. the difference between the dashed and the full curves) are qualitatively similar to the effects of cirrus clouds. However, the absolute magnitude is smaller. It is not small, however, in comparison to solar radiation effects of the cloud cases considered in Fig. 1.

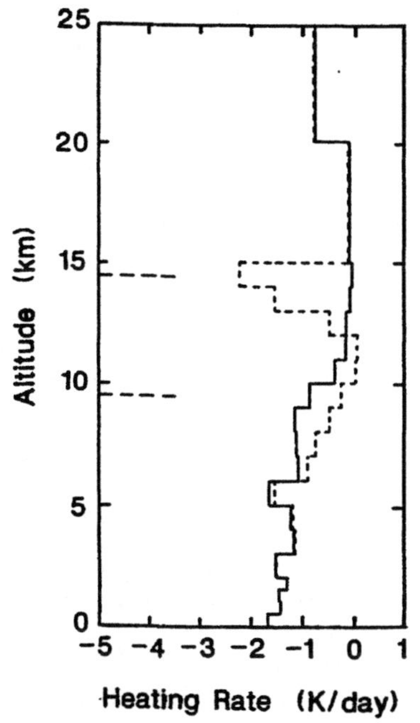

Figure 2. Thermal infrared heating rates for the standard atmosphere (full curve) and for the same case but with water vapour concentration increased up to saturation in the altitude interval from 9.5 to 14.5 km (dashed curve).

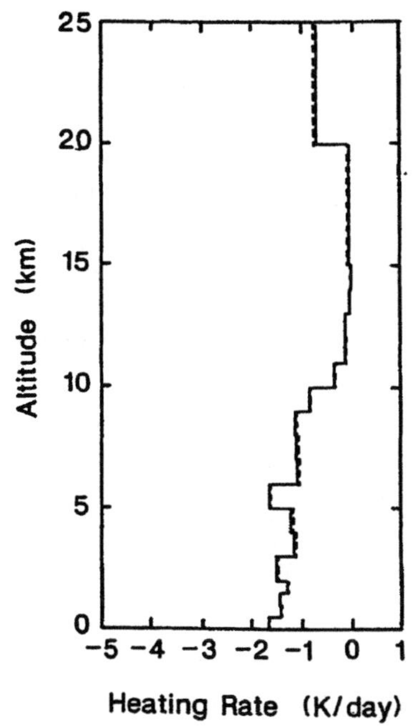

Figure 3. Thermal infrared heating rates for the standard atmosphere (full curve) and for the same case but with doubled CO_2-concentration at all levels.

For comparison, Fig. 3 depicts the infrared heating rate in the standard atmosphere for present and for doubled amount of CO_2, as computed in the same way. The change in heating rate due to doubled CO_2 amounts to about 0.03 K/day at 8 km and -0.7 K/day at 25 km altitude. It is about two orders of magnitude smaller than that due to water vapour saturation in the 5 km interval near the tropopause. However, doubling of CO_2 influences the heating rate at all altitudes, especially in the upper stratosphere. Furthermore, one must note that the expected CO_2-change is a global effect whereas a change of water vapour concentration to the extent postulated above will be achieved, if at all, only very locally.

In any case, these radiation calculations illustrate the potentially strong impact of water vapour and even more of cirrus clouds on radiative heating rates in the upper troposphere.

CONTRAILS OBSERVED ON OCTOBER 18, 1989

On Oct. 18, 1989, an unusually large number of long-living contrails formed over Southern Germany due to exceptional meteorological conditions. This was the second day in a period when from Oct. 17 till Oct. 28 a blocking high was persistent over central Europe with dominant flow at 500 hPa from southwest. The radiosonde observation of Munich of this day at noon-time, see Fig. 4, shows that the temperature decreases down to about 223 K or - 50 ^{0}C at the tropopause which occurs at an altitude of about 12 km (200 hPa in pressure level) at this time. The dew-point-temperature (dashed curve) is considerably below the actual temperature up to about 8 km, indicating dry air in the lower troposphere. In the range above 8 km, the dew-point temperature differs only by about 7 K from the actual temperature which means that the air is rather humid in this range. In fact, as shown in Fig. 5, the relative humidity, which is computed from the sounding data, is of the order 25 % below 8 km. Above 8 km the relative humidity with respect to saturation over water reaches 50 % while with respect to ice saturation it reaches about 80 %. It should be noted that such radiosonde measurements of the dew-point are uncertain up to about 2 K and this corresponds to more than 20 % in humidity (for air temperatures below -30^{0}C) with respect to ice saturation. This error increases for lower temperatures. Hence, it seems well possible that the atmosphere above 8 km was saturated with respect to the ice phase but it was certainly unsaturated for water. The sounding in Fig. 4 shows, moreover, south-westerly winds with 15 to 30 knots in the upper troposphere.

As documented by a series of photos and movie-pictures taken from ground near Oberpfaffenhofen, the sky was clear in the morning of that day. As time went on, during the morning, more and more contrails appeared. The first ones showed up as rather narrow and straight lines. At about noon, the contrails had filled out a large portion of the visible sky. The individual contrails were rather long visible, typically 15 minutes or more. In some cases, the form of the observed ice clouds was deformed and it was not any longer clear whether they were individual sheets of clouds originated from contrails or just natural cirrus clouds. This situation is illustrated by the photo in Fig. 6. Similar observations were made on the day before and on the subsequent days.

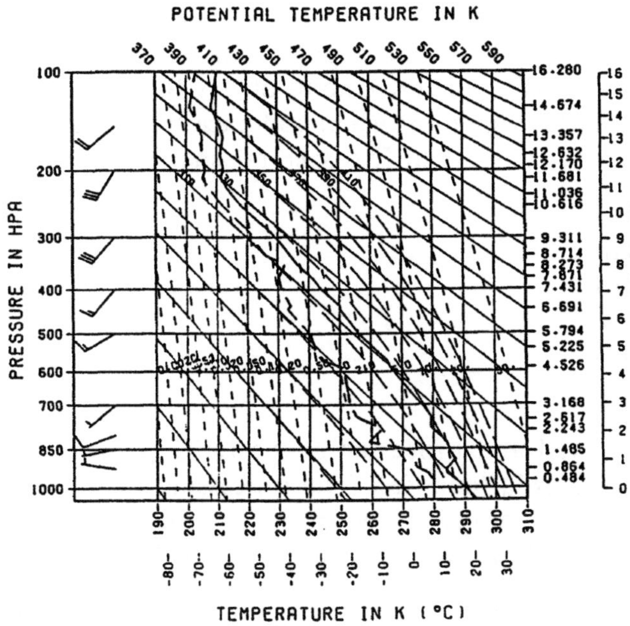

Figure 4. Radiosonde observation for Munich, Oct. 18, 1989, 12.00 UTC, showing wind speed and direction by wind vectors, absolute temperature (full curve) and dew-point temperature (dashed) versus altitude in a Stüve-diagram.

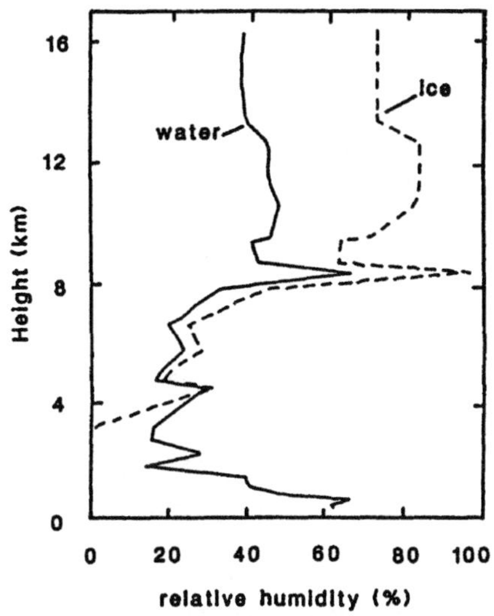

Figure 5. Relative humidity versus altitude relative to saturation over water and over ice.

Figure 6. View from Oberpfaffenhofen towards south-east on Oct. 18, 1989, 12.30 UTC (Photo by H. Höller).

At the time when the photo in Fig. 6 was taken, i.e. at 12:16 UTC (13:16 MEZ), the area was passed over by the NOAA-11 satellite on a polar sun-synchronous orbit at an altitude of about 850 km. This satellite carries among others the Advanced Very High Resolution Radiometer (AVHRR) which measures radiance from the earth with geometric resolution of about 1.1 km at nadir in five spectral channels ranging from the visible to the thermal infrared:

channel 1:	0.58 - 0.68 μm
channel 2:	0.725 - 1.0 μm
channel 3:	3.55 - 3.93 μm
channel 4:	10.3 - 11.3 μm
channel 5:	11.5 - 12.5 μm

The data from this sensor were received and processed at the DFD (German Remote Sensing Data Centre) of DLR in Oberpfaffenhofen. The processing includes computation of reflectances and of brightness temperatures from raw counts of the sensor according to prelaunch and to actual calibration factors.

The data were processed and depicted in Fig. 7 so that ice clouds become clearly visible. For that purpose we make use of the fact (Lee, 1989) that the radiative emissivity of optically thin ice clouds differs strongly in the neighbouring channels 4 and 5. Thin ice clouds are more transparent with respect to infrared radiance from below the clouds near 11μm than near 12μm. Therefore, the difference of the radiation measured in channels 4 and 5 is an effective means to identify thin ice clouds. This effect is well known and earlier has been used by us to analyze cloud heights (Pollinger and Wendling, 1984) as well as various parameters of natural clouds (Kriebel et al., 1989).

For the same scene, we produced a colour figure which has been shown on the conference and which is available from the DLR together with some explanations on a poster "Contrails observed from space", printed in 1989. The colour print is based on an even more complex nonlinear combination of all five AVHRR channels which are projected in three colours such that urban areas or those with only little vegetation appear orange-red, different kinds of vegetation appear ochre to green, coniferous forests are blue-green, water surfaces are blue, snow becomes yellow to white, and clouds are shown in light blue to grey colour. However, for the present discussion, the black and white picture (actually the blue component of the colour print) is sufficient.

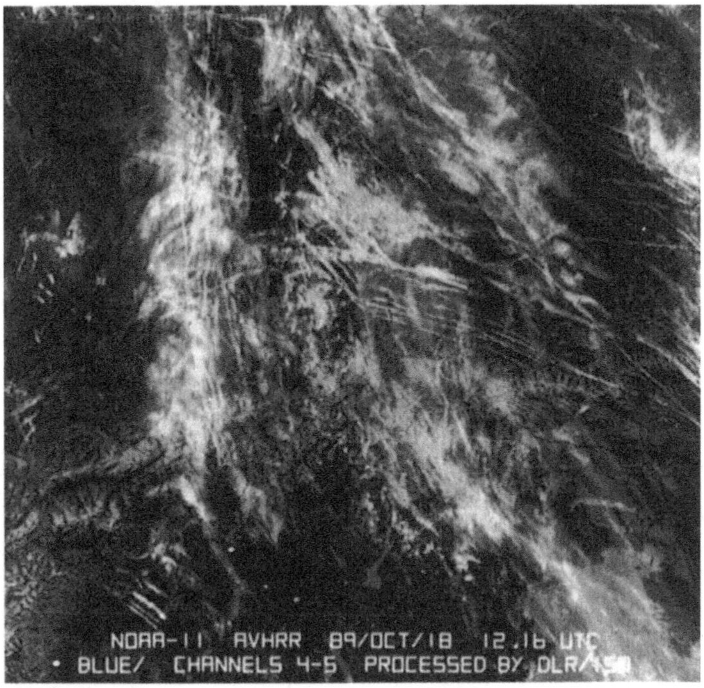

Figure 7. ; of of the NOAA-11-satellite on Oct. 18, 1989, 12.16 UTC in the area between Frankfurt/M. at the upper edge, Venedig and Adria at the lower edge, covering the Alps in the lower third and Kempten nearly in the centre of the picture. Note the white line-clouds which are interpreted as contrails (Processed by DLR/ISM).

Fig. 7 clearly shows that a large portion of the ice-clouds are being formed to narrow bands which very likely are contrails. The following facts support this interpretation:

- The line-forming clouds are aligned with main upper routes of air traffic as plotted in Fig. 8. This is particularly obvious for the route from Villach (VIW) in Austria over Rattenberg (RTT) passing Kempten (KPT) in the direction of the control point TGO and Frankfurt (FFM). Several other routes can be identified.
- At Kempten, this route changes its direction and this is reflected obviously by bends in the line-clouds.
- The lateral distance between individual line-clouds near Kempten is about 5 km. For a wind-speed of 20 knots (10 m/s) this corresponds to a time-difference of about 500 s or 10 minutes, which is a reasonable separation between aircrafts on upper routes.
- The visibility indicates that the clouds are either very thick or sufficiently broad to affect the measured radiance within a single pixel of about 1 km width. The observed line-clouds occasionally reach a width of several pixels, i.e. several kilometers.

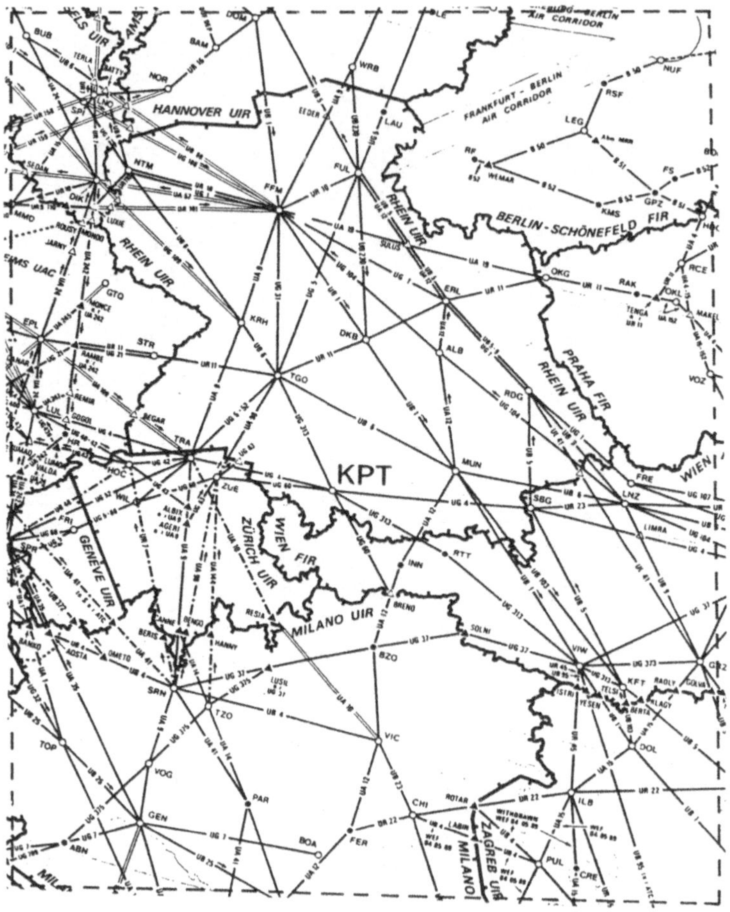

Figure 8. Subsection of the Air Traffic Flow Managment Planning Chart, 8th ed. 1989, showing the upper traffic routes in the same area as taken by the satellite image in Fig. 7.

The last point implies that some of the rather fuzzy cirrus clouds which were observed from ground were in fact wide (up to 10 km) clouds originating from contrails.

On the other hand, the picture shows also large areas filled with rather extended and amorph cirrus clouds. It cannot be decided whether these clouds are the remainders of old contrails or have formed naturally. Hence it is difficult to estimate the fractional area taken by contrails without any further objective criteria.

PATTERN RECOGNITION METHOD

Obviously, an objective method is required to determine which pixel belongs to a contrail. For this purpose, a pattern recognition method has been developed. The method is similar to a method developed by Knöpfle (1988). However, Knöpfle's method does not account for different brightness values of various pixels within cloud areas.

Our method uses a mask, see Fig. 9, of 15 by 15 pixels which is carried over all pixels within the whole matrix of satellite data, i.e. the brightness temperature difference between channels 4 and 5. The centre of the mask is collocated with the pixel under investigation. Within the mask, 16 equidistant directions are distinguished which originate from the centre pixel. Along each direction the consecutive sequence of pixels forms the "mid strip". A pixel of the picture must satisfy the following three conditions before it is classified as being a pixel belonging to a contrail:

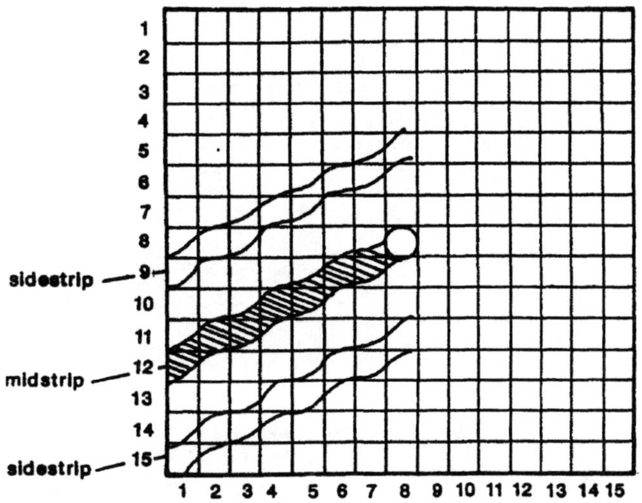

Figure 9. Sketch of the mask used for contrail analysis. Shown is one example of a mid-strip together with two side-strips which are used to identify contrail pixels.

- The pixel must be clouded. I.e., the temperature of the pixel must be lower than a preset threshold value.
- One of the mid-strips along the various predefined directions must be clouded. I.e., at least 5 of 7 pixels along the centre of the strip must be colder than the given limit.
- Side-strips which are parallel to the selected mid-strip but displaced by about 3 pixels sidewards (see Fig. 9) must be cloud-free, i.e. their temperatures must be above a given threshold value.

The threshold values for the mid-strip and the side-strips are functions of the temperature of the centre pixel.

It happens that after one pass over the picture with this procedure some pixels are classified as being contrail pixels, which are obviously not, because they form small isolated entities without linear structure. Therefore, step 2 of the above procedure is applied a second time but now with 56 instead of 16 different radial directions.

Fig. 10 shows for example in the upper part the original data set and in the lower part the processed data set, i.e. the pixels which are classified as being contrails are identified in white. From a subjective basis, the algorithm appears to work successfully. This has been corroborated by applying the same algorithm to further data sets.

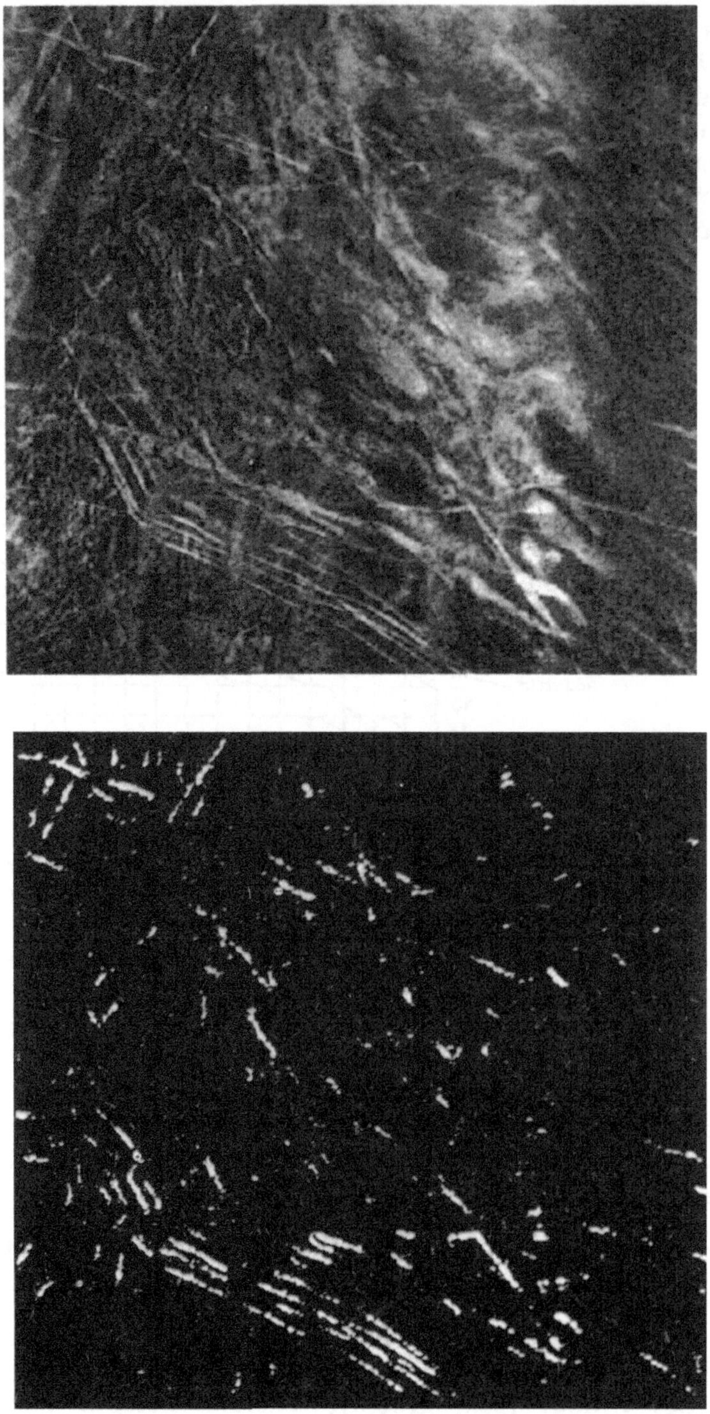

Figure 10. Satellite image as in Fig. 7 (top) together with the result of the contrail analysis method (white pixels in the bottom figure).

In the present case, 1 % of the pixels are classified as belonging to contrails. From a purely subjective point of view we had estimated that up to 8 % of the scene is taken by contrails. Hence, the present algorithm appears to underestimate the actual fraction taken by contrail induced clouds. This clearly shows the difficulty in deciding which pixel is cloudy or not. Improvement is expected by applying the more powerful algorithm package APOLLO (Saunders and Kriebel, 1988) to the data.

STRUCTURE OF A CONTRAIL OBSERVED BY LIDAR

At the same time when the above observations were made from Oberpfaffenhofen, a field-experiment under the heading "International Cirrus Experiment" (ICE) took place over parts of the German Bay. The experiment included observations with a backscatter Lidar ALEX-F (Mörl et al., 1981). The laser pulses were emitted upwards from an aircraft flying at a flight level of 3700 m. The pulse length corresponds to 6 m resolution in altitude. The repetition rate allows for a resolution of about 100 m in flight direction. The measuring aircraft passed below a contrail produced by an airliner a few minutes earlier. Fig. 11 shows a composite of results including the radiosonde observation of Schleswig of that day at 12.00 UTC. The flight measurements were taken at about 13.15 UTC. From the Lidar backscatter signal one can construct contour lines of backscatter intensity. These signals can be taken as a first-order measure of ice-water content. The results in Fig. 11 show that a contrail formed at an altitude of about 8 km. The temperature at the lower edge of the contrail (7.650 km) amounts to about -30 ^{0}C. The contrail extends vertically over about 700 m and laterally over about 3.5 km. Thus the contrail became quite extended although the relative humidity with respect to ice (as deduced from the sounding) amounts to only about 50 %. It is very likely that the sounding underestimates the real humidity considerably. At altitudes between 9 and 10 km, a natural cirrus layer is observed but the measured humidity is still much below saturation. The vertical extent of the contrail is smaller than the horizontal one by about a factor of 5. This corresponds roughly to estimates of the ratio of horizontal to vertical turbulent diffusivities.

By use of the Lidar backscatter signal and a special technique the optical thickness of the contrail shown in Fig. 11 has been determined. This technique makes use of the fact that the Lidar cirrus backscatter signals show remarkable shadowing effects in presence of underlying contrails. Using this shadowing effect the Lidar system can be calibrated to determine extinction coefficients within the contrail with an accuracy of $\pm$20%. The determined optical thickness varies from about 1 at the center of the contrail to about 0.08 at the edge for a wavelength of 1.06 μm. These data have been used to calculate the effects of such a contrail on the radiation balance at the surface. The radiative effects are calculated by using Mie theory and a two stream radiative transfer model. We assume up to now that the ice crystals consist of spheres with radii varying from 2 μm to 40 μm. For an averaged optical thickness of 0.2 (wavelength 0.2 ... 1.0 μm) and a totally cloud covered sky our results show only a small dependence on particle size and the contrail induced cirrus cloud tend to cool the surface at noon by about 6 W/m^2 for a net radiation balance of the clear sky of about 288 W/m^2. This means that for the considered case the contrail induced reduction in the absorbed solar radiation at ground is larger than the reduction of the outgoing longwave flux regardless of particle size.

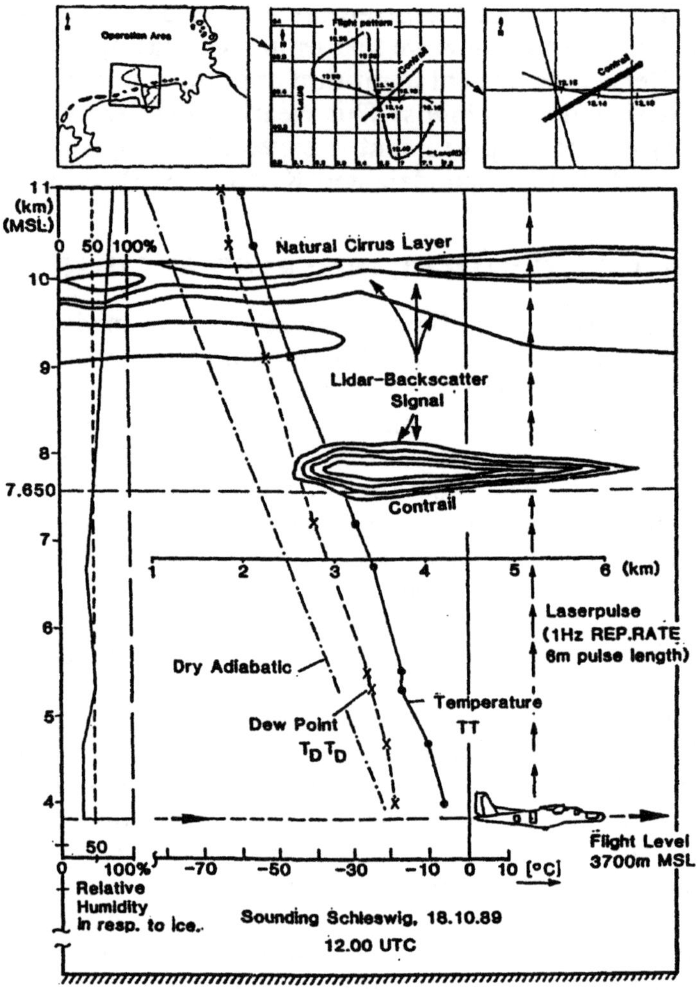

Figure 11. Composite picture showing properties of contrail formation over the northern shore of Germany. The upper small pictures show from left to right: the operation area, the flight pattern, and details of the flight pattern of the measuring airplane and the contrail formed by an airliner. The lower frame contains the results from the radiosonde of Schleswig and the contours of backscatter intensity of the Lidar. The lower edge of the contrail is indicated at altitude 7.650 km. The vertical and horizontal scales of the contour map are equal.

However, definite conclusions regarding the sign of the temperature change at ground cannot be made until the effects of nonspherical ice crystals are considered. Regarding to this difficult and up to now unsolved problem more research is needed in the future.

FURTHER SATELLITE OBSERVATIONS

We have started to evaluate further satellite observations on a routine basis. Up to now, NOAA-AVHRR-data have been processed for 99 days from 8 months within a fixed area of 300000 km^2 between Frankfurt and Genua. At 62 days (63 % of 99) contrails were visible. For 36 of these 99 days, the area taken by contrails was below

0.1 %. For the remaining days, the average contrail coverage was estimated as 1,5 %. The present state of analysis is too preliminary to report final statistics.

Contrails are often observed in the neighbourhood of naturally occuring cirrus clouds. An example of such cases is shown in Fig. 12 from Dec. 20, 1989. Obviously, the width of contrail-induced line clouds grows considerably when approaching areas of natural cirrus clouds.

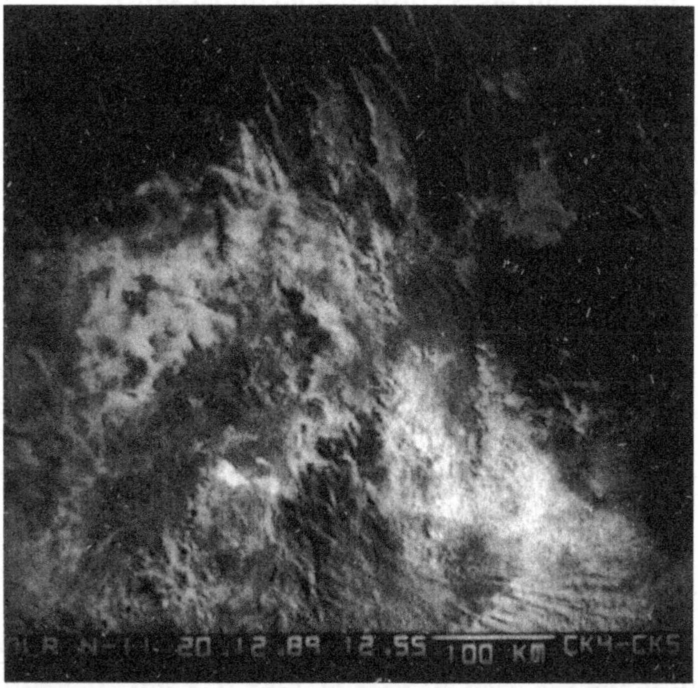

Figure 12. Satellite image similar to that in Fig. 7, but for Dec. 20, 1989, 12.55 UTC. The symbol "m" in the upper right corner represents the location of Munich. Note the many parallel contrails towards north-west at the northern edge of the extended cirrus cloud over the Alps in the middle of the picture.

CONCLUSIONS

Cirrus clouds generated from contrails have the potential to strongly affect radiative heating rates in the atmosphere. The infrared heating or cooling rate magnitude within such cirrus clouds is typically a factor of two larger than that induced by water vapour increase from standard atmosphere to saturation in a 5 km thick layer near the tropopause. There are no reasons to believe that such a change can arise in a realistic manner. In fact, water vapour emissions from air traffic will cause saturation with respect to water only very locally. Nevertheless, in view of the relatively small changes induced by doubling CO_2 the large radiation effects of water vapour and ice clouds are obvious. It still remains to quantify the area and volume fraction in which large increases of water vapour or contrails are caused by airliners.

Contrails may be identified from satellite data of the AVHRR in spite of its rather coarse geometric resolution (about 1 km) by computing the difference of two channels in the thermal infrared. By means of a new pattern recognition method we have a tool to deduce the contrail content of a scene on an objective basis in a repeatable manner so that one can analyse trends and regional distribution. However, this algorithm has yet been applied to a few cases only. By means of Lidar measurements, the spatial dimension of a specific contrail and its optical thickness produced by an airliner have been documented. Although the radiosondes indicate undersaturation with respect to ice and water the contrail takes considerable dimensions.

We have shown that contrails induced by air traffic on upper routes at individual days may cover a considerable portion of the sky. For the specific day of Oct. 18, 1989, which was an extraordinary case, the estimates vary from 1 to 8 % depending on the method used. A climatology is needed to estimate mean cloud covers by contrails within a longer period. Moreover, one still has to perform analysis with radiative transfer and climate models to estimate the effect of such additional (fractional) cloud cover and possible increases in water vapour concentration on the earth's climate. In view of the growing air traffic, such effects will grow with time and may become of significant importance with respect to future climate.

Acknowledgement. This paper is a slightly extended version of that presented by U. Schumann et al. (1990). We thank G. Gesell of the Applied Data Systems Division of the DLR and H. Höller, K.-T. Kriebel, R. Meerkötter, P. Mörl, M. E. Reinhardt, W. Renger, G. Ruppersberg, K.-P. Schickel, and B. Strauß of the Institute of Atmospheric Physics of the DLR for their contributions and help.

REFERENCES

Angell, J. K., 1990: Variation in United States cloudiness and sunshine duration between 1950 and the drought year of 1988. J. Clim. 3, 296-308.

Appleman, H., 1953: The formation of exhaust condensation trails by jet aircraft. Bull. Amer. Meteorol. Soc., 34, 14-20.

Beckwith, W., B., 1972: Future patterns of aircraft operations and fuel burnouts with remarks on contrail formation over the United States. Int. Conf. Aerospace and Aeronatical Meteorol., May 22-26, Washington, D.C., Amer. Meteorol. Soc., Boston, 422-426. pp. 452.

Carleton, A. M. and Lamb, P. J., 1986: Jet contrails and cirrus clouds: a feasibility study employing high-resolution satellite imagery. Bull. Amer. Met. Soc., 67, 301

Changnon, S. A., 1981: Midwestern cloud, sunshine and temperature trends since 1901: possible evidence of jet contrail effects. J. Appl. Meteor., 20, 496-508.

Held, M., 1988: Oekologische Folgen des Flugverkehrs. Tutzinger Materialien, Ev. Akad. Tutzing, Postfach 227, 8132 Tutzing, ISSN 0930-7850, pp. 131.

Joseph, J. H., Levin, Z., Mekler, Y., Ohring, G., Otterman, J., 1975: Study of contrails observed from ERTS 1 satellite imagery. J. Geophys. Res., 80, 366

Knöpfle, W., 1988: Rechnergestützte Detektion linearer Strukturen in digitalen Satellitenbildern. Z. Photogrammetrie und Fernerkundung, 56, 40-47.

Knollenberg, R. G., 1972: Measurements of the growth of the ice budget in a persisting contrail. J. Atmos. Sci., 29, 1367-1374.

Kriebel, K. T., Saunders, R. W., and Gesell, G., 1989: Optical properties of clouds derived from fully cloudy AVHRR pixels. Beitr. Phys. Atmosph. 62, 165-171.

Lee, T. F., 1989: Jet contrail identification using the AVHRR infrared split window. J. Appl. Meteorol., 28, 993-995.

Liou, K.-N., 1986: Influence of cirrus clouds on weather and climate processes: A global perspective. Mon. Wea. Rev., 114, 1167-1199.

Mörl, P., Reinhardt, M. E., Renger, E., and Schellhase, R., 1981: The use of the airborne LIDAR system "ALEX F 1" for aerosol tracing in the lower troposphere. Beitr. Phys. Atmosph., 54, 403-410.

Pfeiffer, M., Fischer, M. (Hrsg.), 1989: Unheil über unseren Köpfen? Flugverkehr auf dem Prüfstand von Ökologie und Sozialverträglichkeit. Quell Verlag, Stuttgart, pp. 239.

Pollinger, W., and Wendling, P., 1984: A bispectral method for the height determination of optically thin ice clouds. Beitr. Phys. Atmosph., 57, 269-289.

Rotter, M., 1987: Auswirkung von Flugzeugkondensstreifen auf die Sonnenbeobachtung und auf die Bewölkungsverhältnisse in Kärnten, Dissertation, Universität Graz.

Saunders, R. W., and Kriebel, K. T., 1988: An improved method for detecting clear sky and cloudy radiances from AVHRR data. Int. J. Remote Sensing, 9, 123-150.

Schumann, U., Gesell, G., Höller, H., Kriebel, K.-T., Meerkötter, R., Mörl, P., Reinhardt, M. E., Renger, W., Schickel, K.-P., Strauß, B. and Wendling, P., 1990: Analysis of air traffic contrails from satellite data - a case study. European Propulsion Forum: Future Civil Engines and the Protection of the Atmosphere, Cologne-Porz, April 3-5, 1990. DGLR-Bericht 90-01, pp. 49-57.

Stephens, G. L., 1989: Cirrus clouds and climate feedback: Is the sky falling and should we go tell the king? FIRE Science Meeting, Monterey, Calif., July 10-14 (sponsored by NASA et al.), p. 327- 331.

Stephens, G. L., and Webster, P. J., 1981: Clouds and climate: Sensitivity of simple systems. J. Atmos. Sci., 38, 235-247.

Weber, G.-R., 1990: Spatial and temporal variation of sunshine in the Federal Republic of Germany. Theor. Appl. Climatol. 41, 1-9.

Wetherald, R. T., and Manabe, S., 1988: Cloud feedback processes in a general circulation model. J. Atmos. Sci., 45, 1397-1415.

AN INVESTIGATION ON THE CLIMATIC EFFECT OF CONTRAIL CIRRUS

K. N. Liou and S. C. Ou

Department of Meteorology/CARSS
University of Utah
Salt Lake City, Utah 84112

G. Koenig

Geophysics Laboratory
Hanscom Air Force Base
Bedford, Massachusetts 01730

ABSTRACT

High cloud cover in the absence of middle and low clouds for Salt Lake City, Utah, during the period 1949-1982 has been analyzed. A significant increase of the mean annual high cloud cover is evident from 1965 to 1982. This increase appears to coincide with the anomalous jet aircraft traffic increase during that period. Analysis of the annual surface temperature for Salt Lake City shows a noticeable increase on the mean annual basis for the period 1965-1982. We have developed a two-dimensional (2-D) cloud-climate model to investigate the perturbation of high cloud cover on the temperature fields. The model consists of a 2-D climate model and a cloud formation model that are interactive through the radiation program. The cloud covers and radiation budgets at the top of the atmosphere computed from the present model compare reasonably well with observations. A 5% uniform increase in high cloud cover at the latitudes between 20° to 70°N would produce an increase in surface temperature by about 1°K with small variations across the latitudes. The positive surface temperature feedback associated with the high cloud cover increase is due to enhanced infrared emission from the additional high cloud cover and specific humidity produced from the 2-D model.

1. INTRODUCTION

Although evidence of changes in global average cloudiness does not exist at present, there have been reports that localized cloudiness has increased. Machta and Carpenter (1971) reported on secular increases in the amount of high cloud cover in the absence of low or middle clouds at a number of stations in the United States between 1948 and 1970. It has been suggested that there may be a link between this increase in cloudiness and the expansion of jet aircraft flights in the upper troposphere and lower stratosphere (Study of Man's Impact on Climate, 1971).

The exhaust plume of jet aircraft, consisting primarily of water vapor, carbon dioxide and some hydrocarbons, produces so-called contrail cirrus. The water vapor within the plume may undergo homogeneous and/or heterogeneous nucleation processes, upon which ice particles form and grow. Contrails may persist only a short time if the ambient air is very dry. In humid conditions, they may persist for minutes to several hours and spread into linear formations a few kilometers in width and tens of kilometers in length. Carleton and Lamb (1986) have used the high-resolution Defense Meteorological Satellite Program (DMSP) imagery to study the possibility of detecting and documenting the occurrence of jet contrails during the period July-November 1979. Six days were selected during which contrails were also identified by surface observations near Champaign, Illinois (Wendland and Semonin, 1982). Based on Carleton and Lamb's studies, contrails tend to occur relatively frequently and they tend to cluster in groups.

Changnon (1981) has carried out a detailed analysis of the cloud observations, sunshine measurements, surface temperatures, and air traffic data over the northern midwest region during the period 1901-1977. There appears to be a downward trend of the annual number of clear days in general, implying that the frequency of cloud formation has been increasing. In particular, high cloudiness increased over north-central Illinois and Indiana during the period 1951-1976, which approximately corresponds to the period of rapid expansion of air traffic. The exact percentage of increase has not been determined, however. Changnon has also reported that the number of months with moderated temperature, defined as the difference between the monthly average maximum temperature and minimum temperature that is less than normal difference, has been increasing. This trend is consistent with the increase in cloudiness.

The heating effects of cirrus clouds on the surface temperature have been investigated by Cox (1971) and determined to be positive or negative depending on the cirrus cloud emissivity. Manabe (1975) has discussed the effect of contrails on the surface temperature and points out the importance of cirrus blackness on the temperature sensitivity experiment. Freeman and Liou (1979) have investigated the increased effect of contrail cirrus cover in mid-latitudes on the radiative budget of the earth-atmosphere system. Extensive one-dimensional numerical experiments have been carried out by Liou and Gebhart (1982) to study the effects of cirrus clouds on equilibrium temperatures. High cirrus clouds above about 8 km would produce a warming effect at the surface. The degree of the warming is a function of the cirrus cloud optical depth (or emissivity). Ou and Liou (1984) have constructed a 2-D climate model based on the energy balance approach to study the effect of the radiative properties of cirrus clouds on global temperature perturbations. Research efforts pertaining to cirrus clouds and climate have been comprehensively reviewed in Liou (1986).

In this paper, we analyze the high cloudiness and surface temperature in Salt Lake City, Utah, from 1948 to 1982. This is described in Section 2. To understand the impact of the potential increase of high cloudiness on temperature, we have carried out a numerical study using a 2-D climate model developed by Ou and Liou (1984). In

Section 3, we present the basic model and the modifications involving a cloud formation program. Perturbation studies and numerical results are presented in Section 4. Concluding remarks are given in Section 5.

2. TRENDS IN HIGH CLOUDINESS AND SURFACE TEMPERATURE FOR SALT LAKE CITY

We have analyzed the high cloudiness (or cloud amount or cloud cover) for Salt Lake City from 1949 to 1982. The three hourly weather observations reported by the National Weather Service at Salt Lake City International Airport were used to determine the sky cover information. For each observation the cloud amount, in tenths of sky coverage, cloud type and visibility were recorded. When the sky cover is less than one tenth for a particular cloud type, it is reported as zero tenths. As many as four cloud types can be reported on a single observation including Ci (high cloud), Ac (middle cloud), and Sc and Cu (low clouds). The high cloud types encountered were cirrus, cirrostratus, cirrocumulus and cirrocumulus standing lenticular.

The monthly average high cloudiness was computed using only those cases in which the sum of the low and middle cloud amount was 0.5 tenths or less. The monthly average high cloudiness was then used to compute the annual high cloudiness. Based on the results of the student t test, the annual high cloudiness, with no low or middle intervening cloudiness, may be considered as two separate populations. The first group corresponds to the annual high cloudiness for the period from 1949 to 1964, while the second accounts for the period 1965 to 1982, as shown in Fig. 1. The mean high cloud amounts for the first and second time periods are 11.8% and 19.6%, respectively. It is noted that Machta and Carpenter (1971) derived a mean high cloud amount of 19.2% for the time period 1965-1969. The mean high cloud amount for the period 1970-1982 was 19.8%, representing only a slight increase over the 1965-1969 period.

We have computed domestic jet fuel consumption from 1956 to 1982. A significant increase in domestic jet fuel consumption in the mid 60's is evident. However, very little increase was shown during the period 1969-1982 (Fig. 1). The sharp anomalous increase in annual high cloudiness appears to coincide with the significant increase in domestic jet fuel consumption in the mid 60's. In fact, the slopes of the linear fit of the annual high cloudiness for the time periods of 1950-1959, 1960-1969 and 1970-1979 are -0.0048, 0.096 and 0.03, respectively. The sharp increase in the domestic jet fuel consumption during the 1960-1969 time frame appears to suggest a possible causative relationship between domestic jet fuel consumption, i.e., increased jet aircraft traffic, and the increase in annual high cloud amount since 1965. During the 1970-1982 time period the frequency of occurrence of high cloud observations decreased. This suggests that the increase in mean annual high cloud amount was due primarily to an increase in cloud amount. Thus, it appears probable that the increase in the high cloud amount could be linked to the increased jet aircraft traffic.

To investigate the potential effect of the increase in high cloud amount on the temperature field, we have analyzed the surface temperature data published by the National Climatic Center. The annual surface temperature values for Salt Lake City over the period 1949-1982 were calculated. The monthly average temperature was obtained by summing the daily maximum and minimum temperatures for the month and dividing by the number of days in the month. The annual surface temperature was obtained by averaging the monthly average temperature information. The temperature data was subdivided into two time periods that coincided with the time periods for high cloud amount, as shown in Fig. 2. The slope for the linear fit of temperature for the first time period is actually negative (-0.032), but the coefficient of determination is fairly small (0.11). This is not the case for the second time period, where the

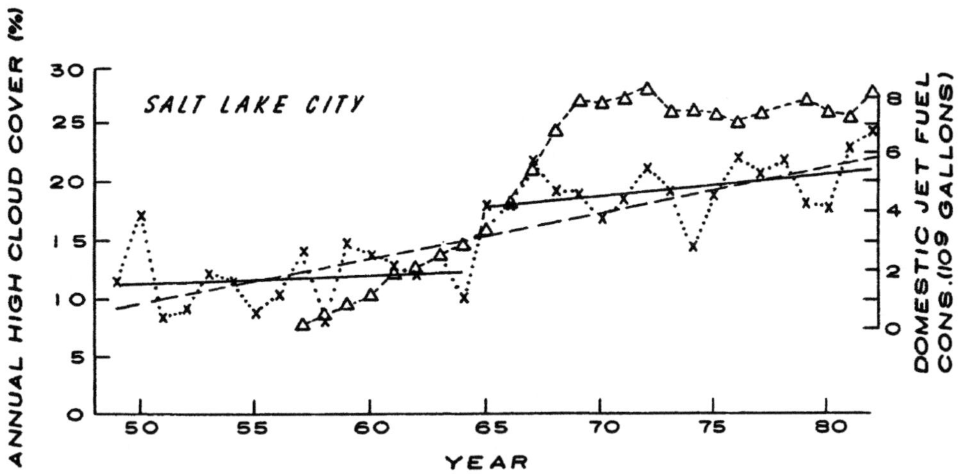

Fig. 1. Mean annual high cloud cover in Salt Lake City from 1948 to 1982 (x) and domestic jet fuel consumption (Δ). The two solid lines are the statistical fitting curves for high cloud cover during the periods 1948-1964 and 1965-1982. The statistical fitting curve for the entire period is denoted by the dashed line.

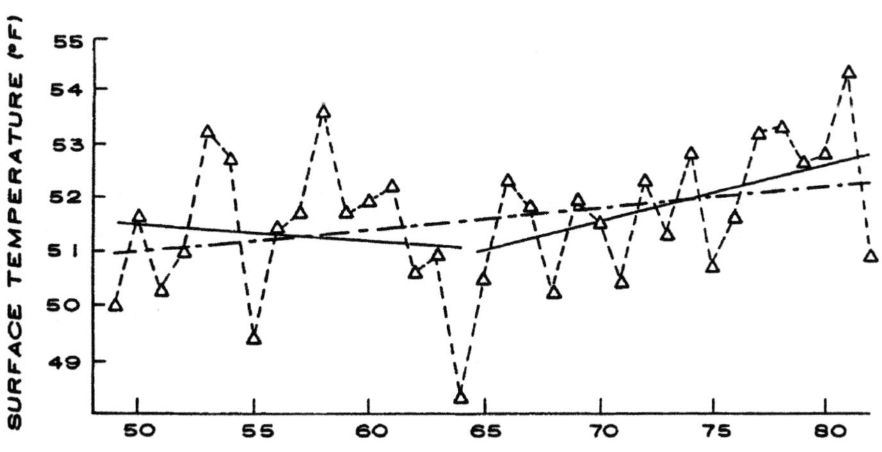

Fig. 2. Mean annual surface temperature in Salt Lake City from 1948 to 1982. The two solid lines are the statistical fitting curves for surface temperature during the periods 1948-1964 and 1965-1982. The statistical fitting curve for the entire period is denoted by the dashed line.

slope is positive (0.107) and the coefficient of determination (0.50) is approximately five times greater than that of the first time period.

An increase in regional surface temperatures could be caused by a number of factors, ranging from natural climatic variabilities to human influence on climate. The correlation between the mean high cloud amounts and mean surface temperatures for the periods 1949–1964 and 1965–1982 appears to suggest that the increase in the surface temperature during the latter period could be related to the increase in high cloud amount.

3. TWO DIMENSIONAL CLOUD-CLIMATE MODEL

To understand the potential effects of high cloud cover variations on the temperature field, we have modified the 2-D climate model developed by Ou and Liou (1984) to perform sensitivity experiments.

a. The 2-D Climate Model

The present 2-D climate model is based on the energy balance between radiative heating and heating produced by horizontal and vertical turbulent transports. The governing equation may be written in the form

$$\frac{\partial}{\partial t}\rho\,[\overline{E}] + \frac{1}{a\cos\lambda}\frac{\partial}{\partial\lambda}\cos\lambda\, F_h + \frac{\partial F_v}{\partial z} = -\frac{\partial F}{\partial z} \quad , \tag{1}$$

where F_h, F_v and F represent the moist static energy fluxes of horizontal and vertical transports, and the net radiative flux, respectively. They are defined by

$$F_h = \rho([\overline{v}]\,[\overline{E}] + [\overline{v'E'}] + [\overline{v^*E^*}]) \quad , \tag{2}$$

$$F_v = \rho([\overline{w}]\,[\overline{E}] + [\overline{w'E'}] + [\overline{w^*E^*}]) \quad , \tag{3}$$

$$F = F_s(\lambda, z) - F_{IR}(\lambda, z) \quad . \tag{4}$$

In the preceding equations, ρ is the air density, t the time, a the earth radius, λ the latitude, and z the height. The moist static energy, $E = C_pT + gz + Lq$, where C_p is the specific heat at constant pressure, g the gravitational acceleration, L latent heat, and q the specific humidity. v and w are the meridional and vertical wind velocities, respectively, and F_s and F_{IR} the net downward solar and IR fluxes, respectively. The symbol $[\overline{\ }]$ represents temporal and zonal averages. The terms on the right hand side (r.h.s.) of Eq. (2) denote the transport of moist static energy by mean motions, transient eddies, and stationary eddies in the meridional direction, respectively. Likewise, the terms on the r.h.s. of Eq. (3) denote their counterparts in the vertical direction. The vertical stationary eddy is a very small quantity. For all practical purposes it may be ignored. The vertical transient eddy may be parameterized using the mixing length theory. The mean flow terms depend on the vertical and horizontal velocities, which are prescribed by using the annual zonal data computed by Oort (1983). This implies that the Hadley circulation is fixed in the model. We have used the mixing length theory to parameterize the large-scale horizontal eddy fluxes in a manner described in Ou and Liou (1984).

b. 2-D Cloud-Moisture Model

An equilibrium 2-D cloud-moisture model has been developed for the simulation of zonally averaged specific humidity, cloud liquid water content (LWC), and cloud cover. The governing equations may be written

$$\frac{1}{a\cos\lambda}\frac{\partial}{\partial\lambda}(\cos\lambda\, F_h^q) + \frac{\partial F_v^q}{\partial z} = -\eta\,\frac{Q_c}{L} + (1-\eta)\,\frac{E_r}{L}\quad, \tag{5}$$

$$\frac{1}{a\cos\lambda}\frac{\partial}{\partial\lambda}(\cos\lambda\, F_h^{\ell}) + \frac{\partial F_v^{\ell}}{\partial z} = \eta\,\frac{Q_c}{L} - P\quad, \tag{6}$$

$$\eta = \frac{h - h_o}{1 - h_o}\quad, \tag{7}$$

where the fluxes of horizontal and vertical transports of specific humidity, q, and cloud LWC, ℓ, are defined by

$$F_h^q = \rho([\bar{v}]\ [\bar{q}] + [\overline{v'q'}] + [\bar{v}^*\ \bar{q}^*])\quad, \tag{8}$$

$$F_v^q = \rho\,([\bar{w}]\ [\bar{q}] + [\overline{w'q'}] + [\bar{w}^*\bar{q}^*])\quad, \tag{9}$$

$$F_h^{\ell} = \rho\,([\bar{v}]\ [\bar{\ell}] + [\overline{v'\ell'}] + [\bar{v}^*\ \bar{\ell}^*])\quad, \tag{10}$$

$$F_v^{\ell} = \rho\,([\bar{w}]\ [\bar{\ell}] + [\overline{w'\ell'}] + [\bar{w}^*\ \bar{\ell}^*])\quad. \tag{11}$$

Parameterizations of the fluxes of the horizontal and vertical transports of specific humidity by eddies follow our previous work described in Ou and Liou (1984). Since global data for the vertical and horizontal transports of cloud LWC by eddies are not available, it is assumed that the ratio of the eddy terms for LWC and specific humidity is approximately proportional to the ratio of the mean LWC and specific humidity. The mean velocity fields are prescribed according to the general circulation statistics provided in Oort (1983).

In Eqs. (5)-(11), Q_c, E_r and P denote the rate of condensation, evaporation of precipitation, and generation of precipitation, respectively; η is the cloud cover, h the relative humidity, and h_o the threshold relative humidity. Equation (7) was derived on the basis of the area-weighted relative humidity in which the relative humidity for the cloud area is taken to be unity. The threshold relative humidity is a closing parameter that is required for the calculation of cloud cover. We follow the parameterized equation suggested by Manabe and Wetherald (1967) in the form

$$h_o = h_{os}\,(p/p_s - 0.02)/0.98 \quad , \tag{12}$$

where p is the atmospheric pressure, p_s the surface pressure, and h_{os} the surface threshold relative humidity, which is obtained from the climatological humidity and temperature values at the surface. At $p \cong 20$ mb, h_o becomes zero. The preceding relative humidity profile was originally proposed for averaged cloudy sky conditions. However, we find that this profile is steeper than the climatological relative humidity profile. Consequently, by using Eq. (12) for clear sky, the averaged cloudy sky relative humidity will be closer to climatological values. It should be noted that the only prescribed value in the model is h_{os}, which is for clear conditions only, and that the model allows clouds to form above the second model layer from the surface.

The derivation of an analytic expression for the rate of condensation, Q_c, requires knowledge of the vertical velocity in the cloudy region. We first define the in-cloud vertical flux of saturation specific humidity, q_s, as $\rho w_s q_s$, where w_s is the in-cloud vertical velocity. Based on the energy conservation principle, Q_c must be proportional to the divergence of $\rho w_s q_s L$. Thus we have

$$\frac{Q_c}{L} = - \frac{\partial}{\partial z}\,(\rho\, w_s\, q_s) \quad . \tag{13}$$

The in-cloud vertical velocity is an unknown parameter. To obtain Q_c, another independent equation is required. Based on the Richardson equation (see, e.g., Dutton, 1976), we have

$$\frac{Q_c}{LC_p} = F_h^T + \frac{w_s}{1-\eta}\,\frac{\partial[\bar{T}]}{\partial z} + \gamma_d \quad , \tag{14}$$

where γ_d is the adiabatic lapse rate, and the flux of the horizontal transport of sensible heat in 2-D space is defined by

$$F_h^T = \frac{1}{a\cos\lambda}\,\frac{\partial}{\partial\lambda}\,\{\cos\lambda\,([\bar{v}]\,[\bar{T}] + [\overline{v'T'}] + [\bar{v}^*\,\bar{T}^*])\} \quad . \tag{15}$$

If $F_h^T < 0$, heating will be produced due to horizontal convergence. However, if $F_h^T > 0$, cooling will be produced. Parameterizations of the fluxes of the horizontal and vertical transports of sensible heat follow the work described in Ou and Liou (1984).

In Eq. (5) the evaporation term associated with precipitation at a given level must be related to the grid-averaged total downward precipitation flux reaching that level and the difference between the specific humidity in the clear column and its saturation value. Thus we have

$$\frac{E_r(z)}{L} = k_r\,\rho(q_s - q_o)\,\tilde{P}(z) \quad , \tag{16}$$

where $k_r = 1.44\ cm^2\ g^{-1}$, which is derived from known microphysical constants, and

$$\tilde{P}(z) = \int_z^{z_t} [\rho\eta P(z') - (1-\eta)\,E_r(z')/L]dz' \quad , \tag{17}$$

where z_t is the cloud top height. Equation (17) represents the accumulated balance between the rates of generation and evaporation of precipitation. Furthermore, the rate of generation of precipitation can be parameterized in terms of the so-called auto-conversion factor and cloud LWC mixing ratio in the form (Ogura and Takahashi, 1971; Sundqvist, 1978)

$$P(z) = \eta\rho\ c_o\ell_c\ [1-\exp(-\ell_o^2/\ell_r^2)] \quad , \tag{18}$$

where $c_o = 10^{-4}\ s^{-1}$ is a coefficient representing the time scale required for cloud particles to become precipitation, and $\ell_r = 5 \times 10^{-4}$ is the mean LWC mixing ratio in precipitation clouds.

Equations (5), (6) and (14) are three partial differential equations consisting of three unknowns: q, ℓ and $\rho w_s q_s$. The solutions of these equations require appropriate boundary conditions. At the surface (z=0), the cloud cover η=0, so that $q(0) = q_s(T_o)\ h_o$, where T_o is the surface temperature and h_o the surface threshold relative humidity. Also, since clouds are not allowed to form at the surface, we have $\ell(0)=0$. The divergence of the in-cloud vertical flux of saturation specific humidity, $\rho w_s q_s$, must be balanced by the surface evaporation and horizontal convergence of moist sources by mean motions and eddies that are parameterized, as described previously. Equations (5), (6), (7), (13), (14), (16) and (18) can then be used to solve for the following variables: q, η, ℓ, P, E_r, Q_c and $\rho w_s q_s$. Once the clouds are formed in model layers, we perform statistical averages to obtain cloud cover and LWC for high, middle and low clouds using the procedures developed by Liou et al. (1985).

c. Radiative Transfer Model

We have used the radiative transfer parameterization program developed by Liou and Ou (1983) and Ou and Liou (1984) in the present study. For infrared radiative transfer, a broadband flux emissivity approach is employed. The entire infrared spectrum is divided into five bands: three for H_2O, one for CO_2, and one for O_3 absorption. The flux emissivity is expressed in terms of a polynomial function of the pressure and temperature corrected path length for H_2O and O_3 based on band-by-band calculations.

For the CO_2 15 µm band, the flux emissivity is parameterized based on the transmittance values derived from line-by-line integrations according to the procedures developed by Ou and Liou (1983). The broadband infrared emissivity, reflectivity, and transmissivity for high clouds are prescribed as functions of the vertically integrated LWC following the parameterization equations developed by Liou and Wittman (1979). Middle and low cloud types are treated as blackbodies in the infrared.

For solar radiative transfer, the spectrum is divided into 25 bands: six for H_2O, one for CO_2 (which overlaps the H_2O 2.7 µm band), and 18 for O_3. The absorptivities for the H_2O and CO_2 bands are calculated according to the method developed in Liou and Sasamori (1975). For O_3, the band absorptivities are computed using the absorption coefficient values complied by Howard et al. (1961). The clear-sky fluxes are determined from known values of the solar zenith angle and absorptivity. For the cloudy case, the broadband solar reflection and transmission values for various cloud types are obtained from parameterized polynomial functions of the integrated cloud LWC and solar zenith angle provided in Liou and Wittman (1979). The net fluxes above and below clouds are then determined based on the functional values of gaseous absorptivities and the cloud radiative properties by means of an interactive procedure that accounts for the effects of cloud-cloud and cloud-surface multiple reflections.

4. EFFECTS OF CONTRAIL CIRRUS ON CLOUD FORMATION, RADIATION BUDGET AND TEMPERATURE

The present 2-D cloud-climate model is a global model that covers both the Northern and Southern Hemispheres. The latitude-height domain is divided into 16 latitudinal belts. There are fourteen 10° belts between 70°S and 70°N and two 20° belts at the poles. The height coordinate consists of 17 layers, including four 50 mb layers between the surface (~ 1000 mb) and 800 mb, seven 100 mb layers between 800 and 100 mb, and six equally divided layers between 100 and 0.1 mb (top level) on the logarithmic scale.

Although the present model consists of two separate models, they are interactive through the radiation program. First, the climatological temperature field is used to drive the cloud model from which cloud covers and LWCs for high, middle and low cloud types are computed. Using the computed cloud data, radiative heating in the atmosphere and surface radiative fluxes are then computed and incorporated in the climate model to obtain the temperature field. The computed temperature field is subsequently fed back to the radiation program and cloud model. Iterations are carried out until the equilibrium state is reached. Generally, only two to three iterations are required.

In the climate model, we use climatological pressure, and air and O_3 densities. The CO_2 concentration is fixed at 330 ppm. Except for the O_3 density, which is taken from values tabulated in McClatchey et al. (1971), all other climatological profiles are taken from data compiled by Oort (1983). The solar constant and duration of sunlight used are 1360 W/m^2 and 12 h (for annual case). The solar zenith angle varies with latitude and can be computed from the equation relating this angle, latitude, solar declination and hour angle. The relative humidity is fixed in the climate model. Thus, the humidity feedback due to temperature perturbations is implicitly included in the model. The surface albedo parameterization has been described in Ou and Liou (1984) and will not be repeated here. This parameterization allows the ice-albedo feedback to occur in the model. Because of the complexity of dynamic transport feedback when the cloud model is interactive with the climate model, this feedback is not considered in the present perturbation study. For this reason, temperature perturbations take place at latitudes where contrail cirrus cover is varied.

a. Control Run

The cloud-climate model is first tuned to the present climate state in terms of the temperature field. Using the procedure developed in Ou and Liou (1984), the model computed temperatures are adjusted to within 0.1°K accuracy of climatological temperatures. Figure 3(a)-(d) show the latitudinal distributions of high, middle, low and total cloud covers computed from the model and climatolology. The climatological cloud values are computed from those derived by London (1957) for the Northern Hemisphere and by Sasamori et al. (1972) for the Southern Hemisphere. Six cloud types were listed by these authors. We have grouped them into high, middle and low cloud types according to the procedure outlined in Liou et al. (1985). In general, the model

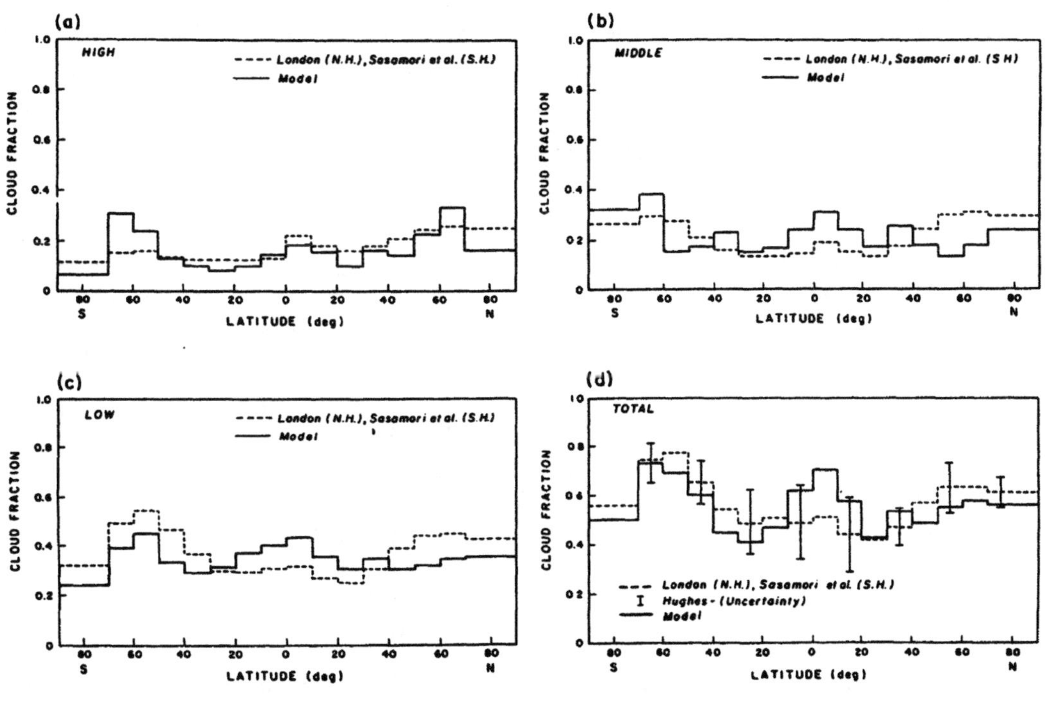

Fig. 3. Latitudinal distributions of high (a), middle (b), low (c), and total (d) cloud covers computed from the present cloud-climate model. Comparisons are made with the cloud covers derived from observed climatological data presented by London (1957) for the Northern Hemisphere and by Sasamori et al. (1972) for the Southern Hemisphere.

produces reasonable cloud cover distributions. High, middle and low cloud covers vary within the range (0.3, 0.1), (0.4, 0.15) and (0.45, 0.25), respectively. Maximum cloud covers ranging from 50-70% are seen in the tropics and mid-latitude storm track regions. In the sub-tropical regions, minimum cloud covers are on the order of 40%. Differences between the model cloud covers and climatological values are generally less than 10%. The best agreement appears to be in mid-latitudes between 30°-60°, where more surface observations were available in the climatological analysis. In the tropics, the model total cloud cover is higher than the climatology indicates. In Fig. 3(d), we have depicted the uncertainty in total cloud cover observations presented by Hughes (1984). It is quite encouraging that the present cloud-climate model produces total cloud covers that are within the observational uncertainty.

Next, we compared the model computed radiation budgets at the top of the atmosphere (TOA) with results derived from satellite observations (Stephens et al., 1981). In Fig. 4, the IR fluxes at TOA are shown (left graph). IR fluxes decrease with increasing latitude, except in the tropics, where convective tower cumulus occur. The model computed IR fluxes in the tropics are higher than observed values. This is due to the fact that the present model does not generate cumulus towers that contain large LWCs. A large difference also occurs in the South Pole regions where the temperature profiles have a large uncertainty. In general, the model IR fluxes are within about 10 W/m^2 of the observed values. Absorbed solar fluxes also compare reasonably well with observations, except in the tropics and Arctic regions (right graph). The latter discrepancies may be due to the solar zenith angle and surface conditions used in model calculations. A number of perturbation experiments were performed to investigate the effects of increasing contrail cirrus in mid-latitudes on the temperature field.

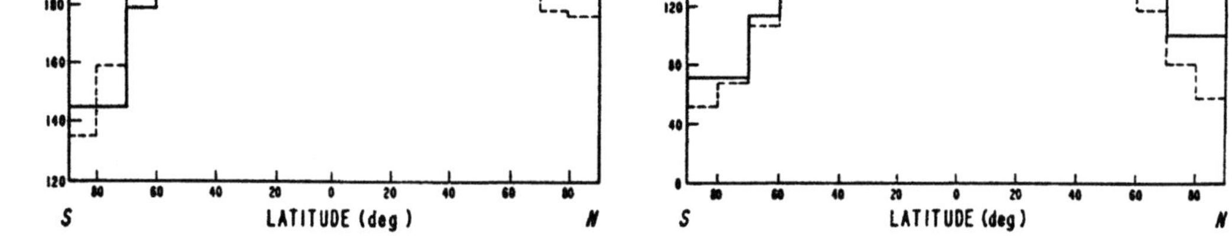

Fig. 4. Latitudinal distributions of net IR flux (left diagram) and net solar flux (right diagram) at the top of the atmosphere computed from the present cloud-climate model. Comparisons are made with satellite derived radiation budgets presented by Stephens et al. (1981).

b. Contrail Cirrus Perturbation Study

We performed numerical experiments to investigate the effect of the increase in contrail cirrus cover on cloud formation and temperature fields. Contrail cirrus cover is increased between 20° to 70°N, roughly corresponding to jet aircraft traffic. Three experiments were carried out, including (1) a 5% increase of high cloud cover with all cloud covers and LWCs fixed in the model (C1); (2) a 5% increase of high cloud cover with interactive cloud covers and LWCs generated from the model (C2); and (3) the same experiment as in (2), except with a 10% increase in high cloud cover (C3). To simplify the perturbation study, the radiative properties of contrail cirrus are assumed to be the same as those of high clouds (Ou and Liou, 1984).

The left graph in Fig. 5 shows the changes in high cloud cover for the region between 20° and 70°N. For C1, high cloud cover is increased by 5% across the latitude, as they are fixed in the numerical experiment. However, for C2, high cloud cover increases by more than 5%, except in the 60-70°N latitude belt. The additional 5% high cloud cover causes both specific humidity and temperature to increase in the atmosphere. The increase in specific humidity is greater in the tropics than in mid-latitudes. Thus, the relative humidity increases more in the tropics, leading to the increase in cloud cover, which is directly calculated from the threshold relative humidity method. Similar results have also been obtained from the perturbation experiment involving doubling of CO_2 concentrations (Ou and Liou, 1987). For an increase in high cloud cover of 10%, the high cloud cover increase would be amplified to a value of about 30% in 20-40°N latitudinal belts as a result of the humidity feedback in the present cloud-

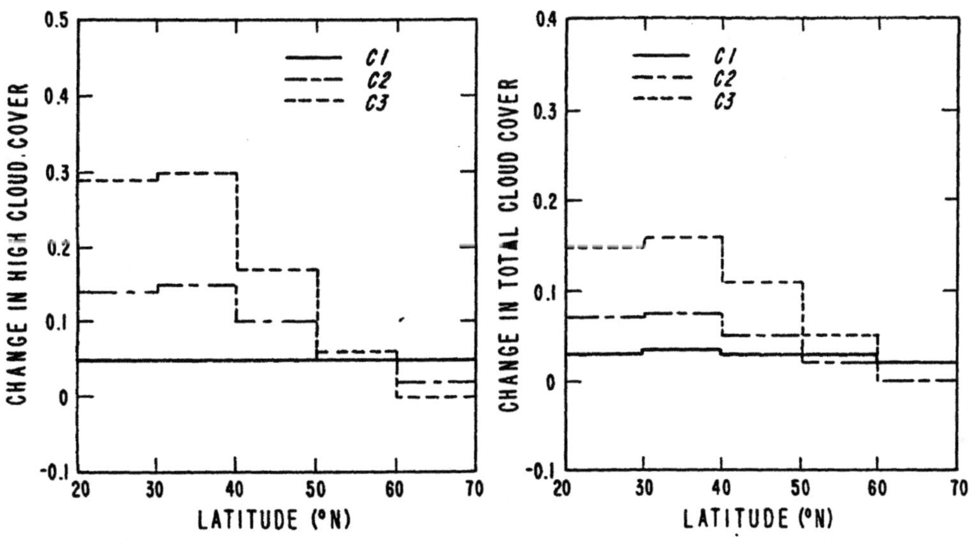

Fig. 5. Changes in high cloud cover (left diagram) and total cloud cover (right diagram) due to the addition of contrail cirrus for the region between 20 and 70°N. C1 denotes a 5% increase in high cloud cover with fixed cloud parameters. C2 denotes a 5% increase in high cloud cover using an interactive cloud program. C3 is the same as C2, except for a 10% increase.

climate model. In terms of the total cloud cover, the increase is less than 5% in the C1 experiment because of cloud overlap effects. In the cases of C2 and C3, increases in the total cloud cover generally mirror those in high cloud cover, except that the magnitudes are smaller due to cloud overlap adjustments.

Changes in atmospheric temperatures for the three perturbation cases are illustrated in Fig. 6. For C1, larger temperature increases of ~6°K at high latitudes are due to the surface albedo feedback included in the model. The amount of temperature increase generally decreases with height. In the case of interactive clouds, the surface albedo feedback effect is substantially reduced. The temperature increase in the C2 case is much smaller than that in C1. Moreover, the maximum temperature increase shifts toward lower latitudes. In Fig. 5, we show that a 5% increase in high cloud cover leads to a substantial amplification in high cloud cover increase (~15%) at 20-40°N, caused by the increase in specific humidity. Enchanced downward thermal infrared emission from additional high clouds is the reason for the temperature increase in the troposphere of low latitudes (IR greenhouse effect). In the cloud perturbation study, low and middle clouds are also increased slightly because of the additional moisture supply. The reduction in the amount of temperature increase in the C2 case is in part due to the increases in low and middle cloud covers (solar albedo effect). In the C3 case, atmospheric temperatures increase more because more high clouds are allowed to form at 20-60°N latitudes. Stratospheric temperatures are only affected slightly due to the high cloud cover increase that occurs in the upper troposphere.

The surface temperature changes produced by the increase of high cloud cover are shown in Fig. 7. In all three cases, surface temperatures increase across the latitudes where cloud cover perturbations take place. For C1, in which clouds are fixed in the model, surface temperature increases from 2 to 6°K with the largest increase occuring at 60-70°N. The surface albedo feedback is the primary reason for this increase, as pointed out previously. When the cloud formation is interactive in the model (C2), the

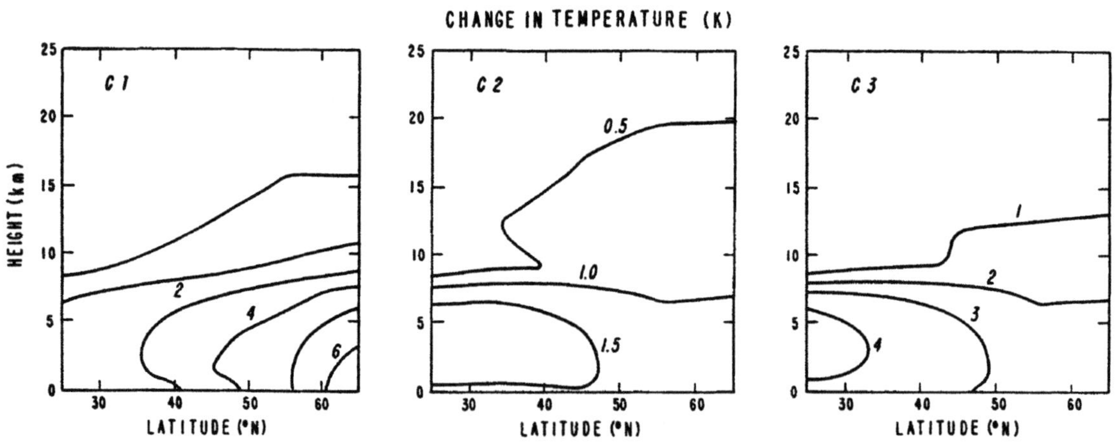

Fig. 6. Changes in atmospheric temperatures due to the increase in high cloud cover for the three cases (C1, C2 and C3) defined in Fig. 5.

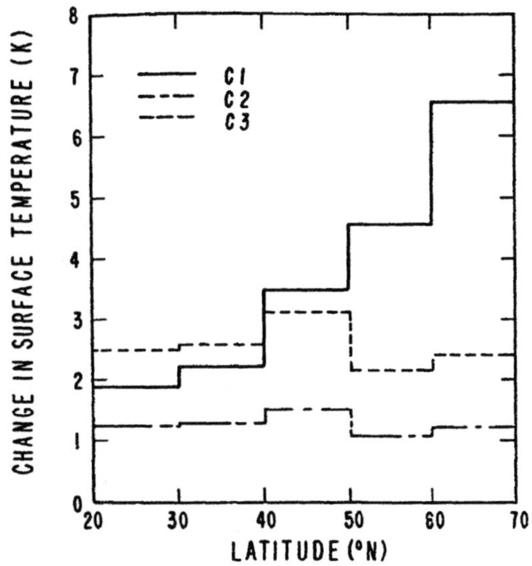

Fig. 7. Changes in surface temperature due to the increase in high cloud cover for the three cases (C1, C2 and C3) defined in Fig. 5.

surface temperature increase is about 1°K but varies with latitude. The decrease in the sensitivity of surface temperature is due to the reduction of surface albedo feedback. In the C3 case, a surface temperature increase of about 2.5°K is obtained because a 10% increase of high cloud cover is used in the perturbation.

In summary, all perturbation experiments involving the increase of high cloud cover reveal increases in atmospheric and surface temperatures. These increases are caused by a positive greenhouse feedback from high cloud cover and specific humidity.

5. CONCLUDING REMARKS

In this paper, we have analyzed high cloud cover in the absence of middle and low clouds for Salt Lake City, Utah, during the period 1949-1982. Based on statistical analyses, the annual high cloud cover may be considered as two separate populations corresponding to the periods 1949-1964 and 1965-1982. A significant increase of mean high cloud cover is evident for the latter period. This increase appears to coincide with the anomalous jet aircraft increase in the mid 60's. An analysis of the mean annual surface temperature for Salt Lake City indicates that the temperature record may also be divided into two groups according to the high cloud cover data. During the period 1965-1982, there is a noticeable temperature increase on the mean annual basis. Factors that could contribute to the increase in regional surface temperatures would be numerous.

The correlation between mean annual high cloud cover and surface temperature in Salt Lake City appears to suggest that the increase in the annual surface temperature during the period 1965-1982 could be caused by the high cloud cover increase associated with jet aircraft traffic during that period.

We have developed a 2-D cloud-climate model to perform sensitivity experiments to understand the perturbation of high cloud cover on cloud-radiation and temperature fields. This model consists of a 2-D climate model and an interactive cloud model that generates cloud covers and LWCs based on the thermodynamic principles. These two models are interactive through the radiation program. The cloud covers and radiation budgets at the top of the atmosphere computed from the cloud-climate model compare reasonably well with observations. Perturbation experiments have been carried out by increasing high cloud cover at the latitudes between 20 to 70°N, roughly corresponding to the jet aircraft traffic corridor. A 5% uniform increase in high cloud cover would produce an increase in surface temperature by about 1°K with some variations across the latitudes. The positive surface temperature feedback produced by the high cloud cover increase is due to enhanced IR emission from the additional high cloud cover and increased specific humidity generated from the 2-D cloud-climate model.

ACKNOWLEDGEMENTS

This research has been supported, in part, by the Air Force Office of Scientific Research Grant AFOSR-87-0294, NASA Grant NAG-1050 and NSF Grant ATM88-15712.

REFERENCES

Carleton, A. M., and P. J. Lamb, 1986: Jet contrails and cirrus cloud: A feasibility study employing high-resolution satellite imagery. *Bull. Amer. Meteor. Soc.*, 67, 301-309.

Changnon, S. A., Jr., 1981: Midwestern cloud, sunshine and temperature trends since 1901: Possible evidence of jet contrail effects. *J. Appl. Meteor.*, 20, 496-508.

Cox, S. K., 1971: Cirrus clouds and climate. *J. Atmos. Sci.*, 28, 1513-1515.

Dutton, J. A., 1976: *The Ceaseless Wind*. McGraw-Hill, 579 pp.

Freeman, K. P., and K. N. Liou, 1979: Climate effects of cirrus clouds. *Adv. Geophys.*, 21, Academic Press, 221-234.

Howard, J. N., J. I. F. King, and P. R. Gast, 1961: Thermal radiation. *Handbook of Geophysics*, MacMillan, Chapter 16.

Hughes, N. A., 1984: Global cloud climatologies: A historical review. *J.Clim. Appl. Meteor.*, 23, 724-751.

Liou, K. N., 1986: Influence of cirrus clouds on weather and climate processes: A global perspective. *Mon. Wea. Rev.*, 114, 1167-1199.

Liou, K. N., and T. Sasamori, 1975: On the transfer of solar radiation in aerosol atmospheres. *J. Atmos. Sci.*, 32, 2166-2177.

Liou, K. N., and G. D. Wittman, 1979: Parameterization of the radiative properties of clouds. *J. Atmos. Sci.*, 36, 1261-1273.

Liou, K. N., and K. L. Gebhart, 1982: Numerical experiments on the thermal equilibrium temperature in cirrus cloudy atmospheres. *J. Meteor. Soc. Japan*, 60, 570-582.

Liou, K. N., and S. C. Ou, 1983: Theory of equilibrium temperatures in radiative-turbulent atmospheres. *J. Atmos. Sci.*, **40**, 214-229.

Liou, K. N., S. C. Ou, and P. J. Lu, 1985: Interactive cloud formation and climatic temperature perturbations. *J. Atmos. Sci.*, 42, 1969-1981.

London, J., 1957: A study of the atmospheric heat balance. New York University, *Final Report*, Contract AF19(122)-166, 99 pp.

Machta, L., and T. Carpenter, 1971: Trends in high cloudiness at Denver and Salt Lake City. *Man's Impact on the Climate* (W. H. Mathews, W. W. Kellogg and G. D. Robinson, Eds.), MIT Press, pp. 410-415.

Manabe, S., 1975: Cloudiness and the radiative convective equilibrium. *The Changing Global Environment* (S.F. Singer, Ed.), Reidel, pp. 175-176.

Manabe, S. and R. T. Wetherald, 1967: Thermal equilibrium of the atmosphere with a given distribution of relative humidity. *J. Atmos. Sci.*, 24, 241-259.

McClatchey, R. A., R. W. Fenn, J. E. A. Selby, F. E. Voltz, and J. S. Garing, 1971: Optical properties of the atmosphere. *AFCRL Environmental Research Papers*, 354, 85 pp.

Ogura, Y., and T. Takahashi, 1971: Numerical simulation of the life cycle of a thunderstorm cell. *Mon. Wea. Rev.*, **99**, 895-911.

Oort, A. H., 1983: *Global Atmospheric Circulation Statistics*, 1958-1973. *NOAA Prof. Paper*, 14, U.S. Department of Commerce, 180 pp.

Ou, S. C., and K. N. Liou, 1984: A two-dimensional radiative-turbulent climate model. I: Sensitivity to cirrus radiative properties. *J. Atmos. Sci.*, 41, 2289-2309.

Ou, S. C., and K. N. Liou, 1987: Effects of interactive cloud cover and liquid water content program on climatic temperature perturbations. *Atmospheric Radiation: Progress and Prospects*, China Science Press, Beijing, China, pp. 443-440.

Sasamori, T., J. London, and D. V. Hoyt, 1972: Radiation budget of the Southern Hemisphere. *Meteorology of the Southern Hemisphere*. *Meteor. Monogr.*, 13, 236 pp.

Study of Man's Impact on Climate, 1971: *Inadvertent Climate Modification*. MIT Press, 308 pp.

Stephens, G. L., G. G. Campbell, and T. H. Vonder Haar, 1981: Earth radiation budgets. *J. Geophys. Res.*, **86**, 9739-9760.

Sundqvist, H., 1978: A parameterization scheme for non-convective condensation including prediction of cloud water content. *Quart. J. Roy. Meteor. Soc.*, **104**, 677-690.

Wendland, W. M., and R. G. Semonin, 1982: Effects of contrail cirrus on surface weather conditions in the midwest - Phase II. *Illinois State Water Survey Contract Report*, **298**, Champaign, 95 pp.

INDEX OF CONTRIBUTORS

GPSR Compliance

The European Union's (EU) General Product Safety Regulation (GPSR) is a set of rules that requires consumer products to be safe and our obligations to ensure this.

If you have any concerns about our products, you can contact us on ProductSafety@springernature.com

In case Publisher is established outside the EU, the EU authorized representative is:

Springer Nature Customer Service Center GmbH
Europaplatz 3
69115 Heidelberg, Germany

Batch number: 10684220

Printed by Printforce, the Netherlands